A Step by Step Approach to the Modeling of Chemical Engineering Processes

Liliane Maria Ferrareso Lona

A Step by Step Approach to the Modeling of Chemical Engineering Processes

Using Excel for Simulation

 Springer

Liliane Maria Ferrareso Lona
School of Chemical Engineering
University of Campinas
Campinas, São Paulo, Brazil

ISBN 978-3-319-88163-8   ISBN 978-3-319-66047-9   (eBook)
https://doi.org/10.1007/978-3-319-66047-9

Printed on acid-free paper

This Springer imprint is published by Springer Nature
The registered company is Springer International Publishing AG
The registered company address is: Gewerbestrasse 11, 6330 Cham, Switzerland

"To Natassia, Alessandra and Jayme"

Preface

The aim of this book is to present the issue of modeling and simulation of chemical engineering processes in a simple, didactic, and friendly way. In order to reach this goal, it was decided to write a book with few pages, simple language, and many illustrations. Sometimes, the rigor of the mathematical nomenclature has been a little simplified or relaxed, to not lose focus on the modeling and simulation. The idea was not to scare readers but to motivate them, making them feel confident and sure they are able to learn how to model and simulate even complex chemical engineering problems. The book is split into two parts: the first one (Chaps. 2, 3, and 4) deals with modeling, and the second (Chaps. 5, 6, and 7) deals with simulation.

To simplify the understanding of how to develop mathematical models, a "recipe" is proposed, which shows how to build a mathematical model step by step. This procedure is applied throughout the entire book, from simpler to more complex problems, progressively increasing the degree of complexity. For each concept of chemical engineering added to the system being modeled (kinetics, reactors, transport phenomena, etc.), a very simple explanation is given about its physical meaning to make the book understandable to students at the start of a chemical engineering course, to students of correlated areas, and even to engineers who have been away from academia for a long time.

The second part of this book is dedicated to simulation, in which mathematical models obtained from the modeling are numerically solved. There are many numerical methods available in the literature for solving the same equations. The focus of this book is not to present all of the existing methods, which can be found in excellent books about numerical methods. In this book, a few effective alternatives are chosen and applied in several practical examples. For each case, the numerical resolution is presented in detail, up to obtaining the final results. The idea is to avoid the reader getting lost in many alternatives of numerical methods, and to focus on how exactly to implement the simulation to obtain the desired results.

When using numerical methods, the simulation step can involve computational packages and programming languages. There are several computational tools for simulation, and it is not possible to say that one is better than another; however, since in most cases a chemical engineering student will work in chemical industries, this book adopts the Excel tool, which is widely used and has a very friendly interface and almost no cost. To develop computational codes, the programming language Visual Basic for Applications (VBA), available in Excel itself, will be used.

It is expected that, with this book, chemical engineering students will feel motivated to solve different practical problems related to chemical industries, knowing they can do so in an easy and fast way, with no need for expensive software.

Campinas, Brazil   Liliane Maria Ferrareso Lona

Organization of the Book

Chapter 1 of the book gives a short introduction and shows the importance of the modeling and simulation issues for a chemical engineer. Important concepts needed to understand the book will also be presented.

Chapter 2 presents a "recipe" (a step-by-step procedure) to be followed to build models for chemical engineering systems, using a very simple problem. The same recipe is used throughout the entire book, to solve more and more complex problems.

Chapter 3 deals with lumped-parameter problems (in steady-state or transient regimes), in which the modeling generates a system of algebraic or ordinary differential equations. The chapter starts by applying the recipe seen in Chap. 2 to simple lumped-parameter problems, but as new concepts of chemical engineering are presented throughout the chapter, the complexity of the problems starts increasing, although the recipe is always followed.

Chapter 4 deals with distributed-parameter systems in steady-state and transient regimes, in which variables such as concentration and temperature change with the position. This kind of problem generates ordinary or partial differential equations. In this chapter, the complexity of examples increases little by little as they are presented, but all of them use the same recipe presented in Chap. 2. In this way, readers can easily understand how to build complex models.

Chapters 5, 6, and 7 are dedicated to numerically solving algebraic equations, ordinary differential equations, and partial differential equations, respectively. There are many different numerical methods available, but in these three chapters a few alternatives will be used because the main purpose of this book is to obtain a fast, robust, and simple way to simulate chemical engineering problems, not to study in detail the different numerical methods available in the literature. All simulations will be done using Excel spreadsheets or codes in VBA.

Chapter 5 uses the Newton–Raphson method to solve nonlinear algebraic equations and presents the concepts of inversion and multiplication of a matrix, available in Excel, to solve linear algebraic equations. Chapter 5 also presents an alternative based on the *Solver* tool available in Excel for both linear and nonlinear algebraic equations. Chapter 6 uses Runge–Kutta methods to solve ordinary differential equations, and Chap. 7 adopts the finite difference method to solve partial differential equations.

I hope this book will be understandable to many people and can motivate all who wish to learn the art of modeling and simulating chemical engineering processes. Good reading!

Acknowledgments

I would like to thank Prof. Maria Aparecida Silva from the Chemical Engineering School at the University of Campinas, who recently retired but, even so, agreed to read the entire book and made valuable corrections and suggestions.

I would also like to thank Prof. Jayme Vaz Junior from the Department of Applied Mathematics at the University of Campinas, who kindly provided the analytical solution shown in Fig. 7.11.

I am very grateful to Prof. Nicolas Spogis who suggested a more didactical way to present one of the subroutines of Chap. 6, and Prof. Roniérik Pioli Vieira, who recommended two examples presented in this book.

I am also deeply grateful to my undergraduate students and teaching assistants, who, in some way or other, made this book better—in particular, João Gabriel Preturlan, Natalia Fachini, and Carolina Machado Di Bisceglie.

Contents

About the Author

Liliane Maria Ferrareso Lona received her bachelor's (1991), master's (1994), and PhD (1996) degrees in chemical engineering from the University of Campinas (Unicamp). She pursued her postdoctoral studies at the Institute for Polymer Research at the University of Waterloo, Canada, from 2001 until 2002. The subject matter in her master's and PhD courses was related to modeling and simulation of petrochemical processes, while her postdoctoral studies focused on the area of modeling, simulation, and optimization of polymerization reactors. In 1996, Liliane became professor at the School of Chemical Engineering–Unicamp, and in 2010 she became full professor with a specialization in the analysis and simulation of chemical processes. Liliane Lona taught for more than 20 times an undergraduate course related to the area of this book. She supervised dozens of grade and undergraduate students in the modeling and simulation area, and many of these works received awards, such as (i) the BRASKEM/Brazilian Association of Chemical Engineering Award (2007), (ii) the Petrobras Award Pipeline Technology (2000 and 2003), and (iii) the Regional Council of Chemistry Award (2000). Liliane published many scientific papers in reputed journals and also served as postgraduate coordinator (2006–2010) and director (2010–2014) of the School of Chemical Engineering–Unicamp.

Chapter 1
Introduction

In chemical engineering, modeling and simulation are important tools for engineers and scientists to better understand the behavior of chemical plants. Modeling and simulation are very useful to design, to scale up and optimize pieces of equipment and chemical plants, for process control, for troubleshooting, for operational fault detection, for training of operators and engineers, for costing and operational planning, etc. A very important characteristic of modeling and simulation is its advantageous cost–benefit ratio because with a virtual chemical plant, obtained from the modeling and simulation, it is possible to predict different scenarios of operations and to test many layouts at almost no cost and in a safe way.

A model can be developed using *deterministic* or *phenomenological* modeling when mathematical equations, based on conservation laws (mass, energy, and momentum balances), are used to represent what physically happens in a system. When conservation laws cannot be applied and an uncertainty principle is introduced, *stochastic* or *probabilistic* models can be used, like population balance or empirical models. This book will address only deterministic or phenomenological models.

A model can be classified as a *lumped-parameter* or *distributed-parameter* model. In a lumped-parameter model, spatial variations in a physical quantity of interest are ignored and the system is considered homogeneous throughout the entire volume. An example of a system that can be modeled using a lumped-parameter model is a perfectly stirred tank, in which variables, such as temperature, concentration, density, etc, are uniform at all points inside the tank, due to the mixing. On the other hand, a distributed-parameter model assumes variations in a physical quantity of interest from one point to another inside the volume. One example of a system that could be modeled using a distributed-parameter model is a tubular reactor, in which the concentration of the reactant decreases along the reactor length. In fact, every real system is distributed; however, if the variations inside the system are very small, they can be ignored and lumped-parameter models can be used. For example, if the agitation in the tank mentioned above was not perfect, small dead zones inside the tank could be generated. However, even so, we

© Springer International Publishing AG 2018
L.M.F. Lona, *A Step by Step Approach to the Modeling of Chemical Engineering Processes*, https://doi.org/10.1007/978-3-319-66047-9_1

could use a lumped-parameter model if we consider—as a *simplifying hypothesis*—perfect agitation, with the small dead zones ignored. Realistic simplifying hypotheses can always be assumed when we are developing models, in order to make them easier to solve.

Another classification used for models is *steady-state* versus *transient* regimes. A system is in a steady state when it does not change over time, which means it is static or stationary. On the other hand, a system is in a transient regime when it changes with respect to time. A transient regime is also called a *non-steady-state*, *unsteady-state*, or *dynamic* regime.

A system modeled by lumped-parameter models is homogeneous and does not present variation throughout the volume, so it is easy to imagine that the final mathematical equation that represents this system (the mathematical model) does not show a derivative with respect to any spatial coordinate. In addition, if this system is in a transient regime (changing over time), the mathematical model must present a derivative with respect to time, while a system in a steady state (static) must not. In this way, it is easy to conclude that a lumped-parameter model in a steady state is represented by *algebraic equations* (AEs), while a lumped-parameter model in a transient regime is represented by *ordinary differential equations* (ODEs).

A distributed-parameter model assumes variation inside the volume, so its mathematical equation (generated from the modeling) will present at least one derivative with respect to spatial coordinates. If the system is in a steady state and there is variation in only one spatial coordinate, the mathematical model will be represented by ODEs, but if this system is in a transient regime, it will be represented by *partial differential equations* (PDEs), with derivatives with respect to time and one spatial coordinate. Finally, if the distributed-parameter model assumes variation in more than one spatial coordinate, it will be represented by PDEs for both steady-state and transient regimes. Fig. 1.1 summarizes all situations analyzed.

Obtaining mathematical equations that represent a system is the *modeling* step. After that, the mathematical equations must be solved. This second part is the *simulation* of the system. The simulation can be done using analytical and numerical methods. This book will focus on numerical solutions.

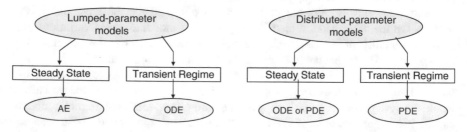

Fig. 1.1 Types of mathematical equations generated from lumped- and distributed-parameter models in steady-state and transient regimes

If the model and simulation are used to predict the behavior of a system that already exists, we say we are doing an *analysis* of the system. On the other hand, if the modeling and simulation are used to define the layout of a system that does not yet exist, we say we are doing *synthesis*.

In this book, Chaps 2, 3, and 4 will focus on how a deterministic mathematical model is developed. Chapter 2 will present a simple recipe that can be used to obtain mathematical models from simple to very complex systems. Chapter 3 will be devoted to lumped-parameter models, and Chap. 4 to distributed-parameter models. Chapters 5, 6, and 7 will address how the mathematical equations generated from the modeling can be solved. Chapters 5, 6, and 7 will focus on numerical solutions for AEs, ODEs, and PDEs, respectively. Despite the huge number of numerical methods available in the literature, this book will focus on just a few numerical methods and will use Excel to solve them. The main idea of this book is to provide a simple and fast tool to obtain numerical solutions for even complex mathematical equations in a targeted and simple way using Excel, which is a very friendly and available tool.

Chapter 2
The Recipe to Build a Mathematical Model

Most chemical engineering students feel a shiver down the spine when they see a set of complex mathematical equations generated from the modeling of a chemical engineering system. This is because they usually do not understand how to achieve this mathematical model, or they do not know how to solve the equations system without spending a lot of time and effort.

Trying to understand how to generate a set of mathematical equations to represent a physical system (to model) and how to solve these equations (to simulate) is not a simple task. A model, most of the time, takes into account all phenomena studied during a chemical engineering course (mass, energy and momentum transfer, chemical reactions, etc.). In the same way, there is a multitude of numerical methods that can be used to solve the same set of equations generated from the modeling, and many different computational languages can be adopted to implement the numerical methods. As a consequence of this comprehensiveness and the combinatorial explosion of possibilities, most books that deal with this subject are very comprehensive, requiring a lot of time and effort to go through the subject.

This book tries to deal with this modeling and simulation issue in a simple, fast, and friendly way, using what you already know or what you can intuitively or easily understand to build a model step by step and, after that, solve it using Excel, a very friendly and widely used tool.

This chapter starts by showing that even if you are a lower undergraduate student, you already known how to do mental calculations to model and simulate simple problems. To prove that, let us imagine a cylindrical tank initially containing 10 m^3 of water. Let us also imagine that the input and output valves in this tank operate at the same volumetric flow rate ($2 \text{ m}^3/\text{h}$), as shown in Fig. 2.1. Assume that the density of water remains constant all the time.

The first question is: 2 h later, what is the volume of water inside the tank? If you say 10 m^3, you are correct. The flow rate that enters the tank is equal to the flow rate that exits ($2 \text{ m}^3/\text{s}$), so the volume of water in the tank remains constant (10 m^3).

© Springer International Publishing AG 2018
L.M.F. Lona, *A Step by Step Approach to the Modeling of Chemical Engineering Processes*, https://doi.org/10.1007/978-3-319-66047-9_2

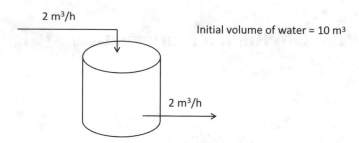

Fig. 2.1 Tank of water with an initial volume equal to 10 m^3

Now, if the input volumetric flow rate changes to 3 m^3/h and the flow rate at the exit remains at 2 m^3/h, what is the volume of water in the tank after 2 h? If you correctly say 12 m^3, it is because you mentally develop a model to represent this tank and after that you simulate it. When the inflow rate becomes 3 m^3/h, by inspection one can conclude easily that the volume of water will increase 1 m^3 in each hour.

Unfortunately, you only know how to do mental modeling and simulation if the problem is very simple. In order to understand how to model and simulate complex systems, let us try to understand what was mentally done in this simple example and transform that into a step-by-step procedure that is robust enough to successfully work also for very complex systems.

2.1   The Recipe

In order to build a mathematical model, three fundamental concepts are used:

1. *Conservation Law*: The conservation law says that what enters the system (E), minus what leaves the system (L), plus what is generated in the system (G), minus what is consumed (C) in the system, is equal to the accumulation in the system (A); or:

$$E - L + G - C = A$$

The accumulation is the variation that occurs in a period of time. This accumulation can be positive or negative, i.e., if what enters plus what is generated in the system is greater than what leaves plus what is consumed in this system, there is a positive accumulation. Otherwise, there is a negative accumulation.

When developing mass and energy balances in the problems presented in this book, we will assume that terms of generation and/or consumption can exist if there are chemical reactions. For example, there is energy generation if there is an exothermic chemical reaction, which will result in an increase in temperature.

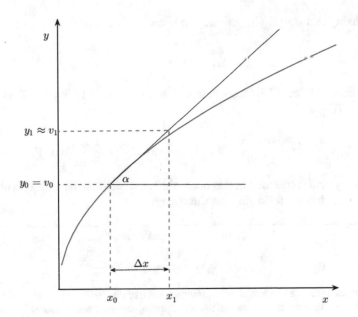

Fig. 2.2 Variation of the dependent variable y with the independent variable x

2. *Control volume*: The control volume is the volume in which the model is developed and the conservation law is applied. All variables (concentration, temperature, density, etc.) have to be uniform inside the control volume. In the example of the tank presented previously, all variables do not change with the position inside the tank (a *lumped-parameter problem*), so the control volume is the entire tank.

3. *Infinitesimal variation of the dependent variable with the independent variable*: Imagine that a dependent variable y varies with x (an independent variable) according to the function shown in Fig. 2.2. Also imagine that in an initial condition x_0 the initial value of y is y_0. To estimate the value of the dependent variable y after an infinitesimal increment in x (Δx), one can draw a tangent line to the curve starting from the point (x_0, y_0), as shown in Fig. 2.2.

The tangent line reaches v_1 at $x = x_1$ ($x_1 = x_0 + \Delta x$). If the increment Δx is sufficiently small, it follows that $y_1 \cong v_1$, and it is possible to obtain the value of y_1 using the concept tangent of α:

$$\tan \alpha = \frac{y_1 - y_0}{x_1 - x_0} = \left.\frac{dy}{dx}\right|_{x_0, y_0}$$

so:

$$y_1 = y_0 + \Delta x \left.\frac{dy}{dx}\right|_{x_0, y_0}$$

Generalizing and simplifying the way to show the index of the derivative:

$$y_{i+1} = y_i + \frac{dy_i}{dx}\Delta x \tag{2.1}$$

Equation (2.1) could be also obtained using the first term of a Taylor series expansion (Eq. 2.2):

$$y_{i+1} \cong y_i + \frac{dy_i}{dx}\Delta x + \frac{1}{2!}\frac{d^2 y_i}{dx^2}(\Delta x)^2 + \frac{1}{3!}\frac{d^3 y_i}{dx^3}(\Delta x)^3 + \frac{1}{4!}\frac{d^4 y_i}{dx^4}(\Delta x)^4 + \cdots \tag{2.2}$$

For all systems presented in this book, the same recipe will be used to obtain the mathematical model, following the three steps:

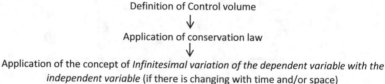

Definition of Control volume

↓

Application of conservation law

↓

Application of the concept of *Infinitesimal variation of the dependent variable with the independent variable* (if there is changing with time and/or space)

2.2 The Recipe Applied to a Simple System

Keeping in mind the three fundamental concepts presented in Sect. 2.1, let us apply the step-by-step procedure (the recipe) to model the tank presented previously. This procedure, used to model this simple system, will be the same used throughout the entire book, in order to solve more and more complex problems.

As stated in Sect. 2.1, the entire tank must be considered as the *control volume* because we are dealing with a lumped-parameter problem. The *dashed line* in Fig. 2.3 shows the control volume considered in this case.

Fig. 2.3 Tank of water with the control volume used in the modeling

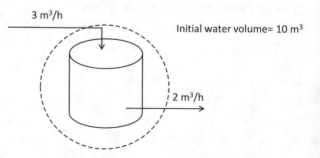

The application of the *conservation law* to the *control volume* yields the expression presented by Eq. (2.3) (observe that there is neither generation nor consumption of water):

$$E - L = A \qquad (2.3)$$

The E and L terms can be easily obtained, since the flow rates that enter and leave the tank are known (3 m³/h and 2 m³/h, respectively); however, how can the accumulation term be obtained?

In order to obtain the accumulation term, we can use the concept of the *infinitesimal variation of the dependent variable with the independent variable*. So if we say that at a time t the mass of water in the tank is M (kg), after an infinitesimal period of time (Δt) the mass of water in the tank will be $M + \frac{dM}{dt} \Delta t$ (kg) (see analogy with Eq. (2.1)). The table below summarizes this information.

t	$t + \Delta t$	Dimension
M	$M + \dfrac{dM}{dt} \Delta t$	kg

The amount of water accumulated in the tank in a period of time Δt is the mass of water at the time $t + \Delta t$ minus the mass of water at the time t, so the accumulation term (A) is given by:

$$A = M + \frac{dM}{dt} \Delta t - M$$

or:

$$A = \frac{dM}{dt} \Delta t \quad (\text{kg})$$

Since the mass is the density times the volume ($M = \rho V$) and the density remains constant, the accumulation term can also be written as:

$$A = \rho \frac{dV}{dt} \Delta t \quad (\text{kg})$$

A very important tool to check if a model is correct is to do a dimensional analysis on all terms of the conservation law equation.

If we calculate how much water accumulates in the tank in a period of time Δt, we have to consider how much water enters and leaves the tank in this same interval of time (Δt). So, in a period of time Δt, the amount of water that enters and leaves the tank is:

$$E = 3(\text{m}^3/\text{h}) \, \rho(\text{kg/m}^3) \, \Delta t(\text{h}) \quad \rightarrow \quad E = 3\rho \, \Delta t(\text{kg})$$
$$L = 2(\text{m}^3/\text{h}) \, \rho(\text{kg/m}^3) \, \Delta t(\text{h}) \quad \rightarrow \quad L = 2\rho \, \Delta t(\text{kg})$$

so applying the conservation law for the period of time Δt yields:

$$\underbrace{3\rho\Delta t(\text{kg})}_{\text{Enters (E)}} - \underbrace{2\rho\Delta t(\text{kg})}_{\text{Leaves (L)}} = \underbrace{\rho\frac{dV}{dt}\Delta t(\text{kg})}_{\text{Accumulation (A)}} \tag{2.4}$$

Observe that the density (ρ) is present in the three terms of the mass balance, so Eq. (2.4) can be simplified. In this way, we can conclude that when the density remains constant, we can directly do the volume balance (instead of mass balance). In this case, the accumulation term, as well as the terms E and L, could be obtained as shown below:

t	$t + \Delta t$	Accumulation	Dimension
V	$V + \dfrac{dV}{dt}\Delta t$	$\dfrac{dV}{dt}\Delta t$	m^3

$$E = 3(\text{m}^3/\text{h})\Delta t(\text{h}) \quad \rightarrow \quad E = 3\Delta t(\text{m}^3)$$
$$L = 2(\text{m}^3/\text{h})\Delta t(\text{h}) \quad \rightarrow \quad L = 2\Delta t(\text{m}^3)$$

so the balance becomes:

$$\underbrace{3\Delta t(\text{m}^3)}_{\text{Enters (E)}} - \underbrace{2\Delta t(\text{m}^3)}_{\text{Leaves (L)}} = \underbrace{\frac{dV}{dt}\Delta t(\text{m}^3)}_{\text{Accumulation (A)}} \tag{2.5}$$

Observe that Eqs. (2.4) and (2.5) are the same, and after simplifying terms this yields:

$$\frac{dV}{dt} = 1 \tag{2.6}$$

Equation (2.6) represents the model for this simple system and agrees with the mental calculation you did previously. Having completed the modeling stage, we need to do the simulation, which is nothing more than solving, by analytical or numerical methods, the equations generated from the modeling. In our case, as the system is greatly simplified, a single and very simple ordinary differential equation (ODE) is generated from the modeling, and it will be solved by direct integration.

To solve this ODE, one initial condition is necessary. In our case, we know that in the beginning of the operation, the volume of water in the tank is 10 m^3. So the initial condition is:

$$\text{At} \quad t = 0, \quad V = 10 \text{ m}^3$$

Solving Eq. (2.6) using the initial conditions yields:

$$V = 10 + t \tag{2.7}$$

Equation (2.7) shows how the volume of liquid in the tank varies with time, making it possible to predict, for example, the time it takes for the liquid to overflow the tank (also observe that the equation says that after 2 h, the volume of water is 12 m^3, as predicted previously).

The procedure adopted for this simple example will be used from now on for more and more complex examples.

Proposed Problem

2.1) Develop a model for the tank presented in Fig. 2.3, but consider that the flow rate of water that leaves the tank (Q_{out}, m^3/h) depends on the level of the water (h) inside the tank, in the way $Q_{out} = 1 + 0.1h$ (m^3/h). This can be a real situation because as the column of water increases, the pressure on the exit point also increases, and consequently the exit flow rate becomes greater. Assuming that the initial volume of water inside the tank is equal to 10 m^3 and the cross-sectional area of this tank is equal to 1 m^2, the initial level of water (h) is 10 m, so in the beginning, the flow rate that leaves the tank (Q_{out}) is equal to 2 m^3/h. In the beginning, the input flow rate is equal to 2 m^3/h, so the volume of water remains constant, in a steady-state regime. If for some reason the inflow rate varies from 2 to 3 m^3/h, develop a mathematical model to represent how the level of water inside the tank varies with time. Define the initial condition needed to solve the equation generated from the modeling.

Chapter 3
Lumped-Parameter Models

This chapter uses the recipe presented in Chap. 2 to develop models for different systems related to chemical engineering. The examples presented in this chapter deal with lumped-parameter problems, in which spacial variations in a physical quantity of interest are ignored. As shown in Fig. 1.1, lumped-parameter problems in a steady state are represented by algebraic equations, and, in a transient regime, by ordinary differential equations. In this chapter, we will only develop mathematical models using the recipe presented in Chap. 2. Numerical solution (using Excel) of algebraic and ordinary differential equations will be seen in Chaps. 5 and 6, respectively.

As mentioned in Chap. 1, one example of a lumped-parameter problem is a perfectly stirred tank, in which we assume that the agitation is so perfect that the system can be considered homogeneous (no internal profiles of concentration, temperature, etc).

Section 3.1 will present three introductory examples of lumped-parameter modeling involving mass, energy, and volume balances. Sections 3.2 and 3.3 will revisit some concepts about heat transfer and chemical reactions, needed to model problems with a somewhat greater complexity level, and will show five practical examples of how to model systems involving these concepts.

3.1 Some Introductory Examples

This section will be presented in the form of three introductory examples, which will explore mass, energy, and volume balances.

Example 3.1 Mass Balance in a Perfectly Stirred Tank
Let us consider a perfectly stirred tank initially containing 10 m^3 of pure water. Assume that the tank contains inlet and outlet valves, both operating at the same flow rate (2 m^3/h), so the volume of water inside the tank does not change over time

© Springer International Publishing AG 2018
L.M.F. Lona, *A Step by Step Approach to the Modeling of Chemical Engineering Processes*, https://doi.org/10.1007/978-3-319-66047-9_3

Fig. 3.1 Perfectly stirred
tank being fed with a NaOH
solution

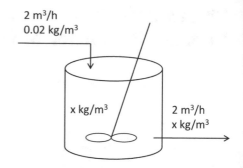

(assuming an incompressible fluid, i.e., constant density). In the beginning, the inlet
stream contains just water. At some point, a solution of NaOH at a concentration of
0.02 kg/m^3 is fed instead, at the same flow rate (2 m^3/h). What is the concentration
of NaOH in the liquid leaving the tank?

Solution:

By inspection, one can imagine that the concentration of NaOH in the tank is
initially zero (pure water) and when the solution of NaOH starts being fed,
the concentration of NaOH in the tank starts increasing, but it does not exceed
0.02 kg/m^3.

One can also imagine that, if the agitation is perfect, the concentration of NaOH
at all points inside the tank is the same, including the point very close to the outlet
valve, so we can conclude that the concentration of NaOH inside the tank is equal to
the concentration of NaOH that leaves the tank. As we do not know the value of this
concentration (and remember, it will change over time), we will assume its value is
equal to x (kg/m^3).

As this is a lumped-parameter problem, the entire tank must be considered as the
control volume. A scheme that represents our problem, from the point at which a
solution of NaOH starts being fed, can be seen in Fig. 3.1.

Let us start doing the mass balance of NaOH inside the tank (the control
volume). As there is no generation or consumption of NaOH (no chemical reaction), the *conservation law* applied to this case yields $E - L = A$.

The accumulation term can be obtained by the concept of the *infinitesimal
variation of the dependent variable with the independent variable*, considering
the amount of NaOH at time t and at time $t + \Delta t$. The amount of NaOH in the
tank is the concentration (x) of NaOH in the tank (kg/m^3) multiplied by the volume
(V) of the tank (m^3).

t	$t + \Delta t$	Dimension
Vx	$Vx + \dfrac{d(Vx)}{dt}\Delta t$	kg

The amount of NaOH that accumulates in a period Δt is the amount of NaOH at
time $t + \Delta t$ minus the amount of NaOH at time t.

$$A = \frac{d(Vx)}{dt} \Delta t \ \text{(kg)}$$

All terms of the conservation law (the amounts of NaOH that enter, leave, and accumulate) must be considered in the same period of time, in this case, Δt (h).

The amount of NaOH (kg) that enters the tank in Δt (h) can be obtained by multiplying the volumetric inflow rate (2 m^3/h), the inflow concentration (0.02 kg/m^3), and the period of time Δt (h):

$$E = 2\left(\frac{m^3}{h}\right) 0.02 \left(\frac{kg}{m^3}\right) \Delta t(\text{h})$$

$$E = 0.04 \ \Delta t \ \text{(kg)}$$

In the same way, one can obtain the amount of NaOH that leaves the tank in the same period of time Δt (h):

$$L = 2\left(\frac{m^3}{h}\right) x \left(\frac{kg}{m^3}\right) \Delta t(\text{h})$$

$$L = 2 x \Delta t \ \text{(kg)}$$

The mass balance can be obtained by substituting the terms E, L, and A in the conservation law:

$$0.04\Delta t - 2x\Delta t = \frac{d(Vx)}{dt} \Delta t \qquad (3.1)$$

Observe that all terms have the same dimension (kg). Simplifying the Δt term yields:

$$\boxed{\frac{d(Vx)}{dt} = 0.04 - 2x}$$

In our case, the volume of the tank remains constant and is equal to 10 m^3, so the final equation to represent the concentration of NaOH in the tank (and leaving the tank) is:

$$\frac{dx}{dt} = \frac{0.04 - 2x}{10}$$

This ordinary differential equation (ODE) is the mathematical model that represents the stirred tank. The simulation of this system is obtained by solving this ODE analytically or numerically. In order to solve this ODE, an initial condition is needed. In our problem the initial condition available is: at $t = 0$ h, $x = 0$ kg/m^3 (pure water).

After solving the ODE, one can obtain a profile of the concentration of NaOH inside the tank over time, as shown in Fig. 3.2. As mentioned before, the NaOH

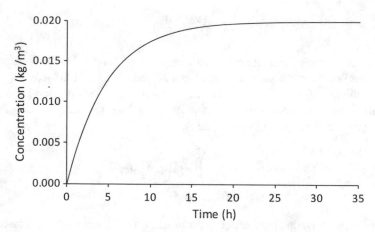

Fig. 3.2 Concentration profile of NaOH leaving the tank over time

concentration indeed starts at zero and tends toward the value 0.02 kg/m^3, which is the concentration of the stream fed into the tank. When the concentration of NaOH inside the tank does not vary anymore with time, we say the system reaches a *steady state*. So, in our case, the system is in a *transient state* (when the concentration of NaOH changes with time) and then reaches a *steady state* (when the concentration of NaOH remains constant over time).

We can also obtain the NaOH concentration inside the tank in a steady state directly (with no need to draw a graph) by setting a value of the accumulation term in the conservation law equal to zero ($E - L = 0$). This is possible because in a steady state the concentration of NaOH stays the same over time. In this way, the mass balance becomes (compare this with Eq. 3.1):

$$0.04 - 2x = 0$$

This equation is easily solved and yields $x = 0.02 \text{ kg/m}^3$ (as expected; see Fig. 3.2).

It is important to observe that some systems do not reach a steady state. Note that in the example presented in Chap. 2, the volume of the liquid inside the tank will increase indefinitely with time, until the tank overflows.

In the next example, there will be a small increase in complexity because the two examples previously presented will be combined (changes in volume and in concentration), and the system will be represented by two ODEs.

Example 3.2 Mass and Volume Balance in a Perfectly Stirred Tank
Assume now a situation in which there are variations with time of both volume and NaOH concentration. In this case, the models developed in Chap. 2 and in Example 3.1 have to be combined. The system can be represented by Fig. 3.3. At the beginning, a perfectly stirred tank contains 10 m^3 of pure water. Shortly thereafter, the tank starts to be fed at a rate of $3 \text{ m}^3/\text{h}$ with a NaOH solution at a concentration of 0.02 kg/m^3. Simultaneously, an outlet valve is opened, allowing the fluid to leave

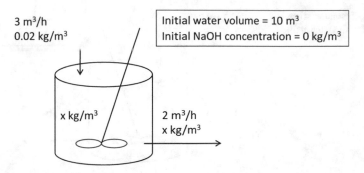

Fig. 3.3 Perfectly stirred tank with variations in NaOH concentration and volume

the tank at a rate of 2 m³/h. How do the volume and the concentration of NaOH vary with time inside the tank?

Solution:

Chapter 2 develops the volume balance for this tank and obtains:

$$\frac{dV}{dt} = 1$$

Developing the mass balance of NaOH for the tank, as per Example 3.1, we obtain:

$$\frac{d(Vx)}{dt} = 0.06 - 2x$$

If the volume changes with time, this yields:

$$\frac{d(Vx)}{dt} = V\frac{dx}{dt} + x\frac{dV}{dt}$$

So the equation system that represents this tank is:

$$\frac{dV}{dt} = 1$$

$$\frac{dx}{dt} = \frac{0.06 - 3x}{V}$$

and the initial conditions are: At $t = 0$, $V = 10$ m³ and $x = 0$ kg/m³.

Both equations have to be solved simultaneously to generate profiles of the volume and NaOH concentration inside the tank over time.

Example 3.3 Energy Balance in an Insulated Stirred Tank

Now let us take a step further by considering a simple energy balance. Variations in temperature in chemical plants can occur basically due to generation or consumption of energy as a consequence of exothermic or endothermic chemical reactions,

Fig. 3.4 Insulated stirred tank fed with water at 50 °C

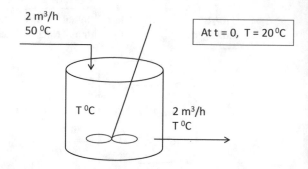

and due to heat transfer phenomena, such as radiation, conduction, and natural or forced convection. In this simple example, the variation in temperature will occur only because the system is fed with a fluid at high temperature.

Imagine an insulated, perfectly stirred tank containing 10 m³ of water at 20 °C. At some point the inlet and outlet valves are opened, both operating at a flow rate of 2 m³/h. Assume that the density of the water remains constant, even if the temperature varies, so the volume inside the tank remains constant and equal to 10 m³. If the temperature of the water fed into the system is 50 °C, how does the water temperature inside the tank change over time? How long does it take to reach a steady state? What is the temperature at the steady state?

Solution:

By inspection, one can imagine that initially the temperature of the water inside the tank is 20 °C and increases until it reaches 50 °C (observe that this tank is insulated and does not lose heat to the environment). Figure 3.4 shows the proposed system.

As the water in the tank is perfectly mixed, the entire tank has to be considered as the control volume. There is no generation or consumption of energy inside the tank, so the conservation law applied to this system yields:

$$E - L = A$$

Initially we will analyze the accumulation of energy inside the tank in a period of time Δt. As we are dealing with the energy balance, the amount of accumulated energy must be given in units of energy, such as joules (J), BTU, cal, etc. Analyzing our problem, we will try to infer how to represent this energy using both dimensional analysis and the physical meaning of the variables.

The amount of heat accumulated in the system depends on (i) the amount of material, given by its mass (the greater the mass, the greater the amount of accumulated heat); (ii) the temperature (i.e., the higher the temperature is, the more energy the material holds); and (iii) the characteristics of the material (i.e., its ability to accumulate heat, given by its specific heat). Thus, the amount of energy in the system at a given time can be represented (using international units) by:

$$M = \text{mass (kg)}$$

$Mc_pT\,(\text{J})$ $$c_p = \text{specific heat } (\text{J/kg}^\circ\text{C})$$

$$T = \text{temperature}(^\circ\text{C})$$

The mass (M) can be represented by the product of the volume and density $(M = \rho V)$. Hence, the amount of accumulated heat in a period of time Δt is given by the amount of heat at time $t + \Delta t$ minus the amount of heat at the previous time t, as can be seen below:

t	$t + \Delta t$	Dimension
$V\rho c_p T$	$V\rho c_p T + \dfrac{d(V\rho c_p T)}{dt}\Delta t$	J

$$A = \frac{d(V\rho c_p T)}{dt}\Delta t \quad (\text{J})$$

When developing a model, simplifying assumptions can be considered in order to make the simulation easier. In our example, one can assume that the density and the specific heat of the water do not vary over time, even if the temperature changes. Doing so, the accumulation term becomes:

$$\text{Accumulation} = A - V\rho c_p \frac{dT}{dt}\Delta t \quad (\text{J})$$

The next step is to analyze the input and output terms of the conservation law. We need to obtain the amounts of energy (in joules—the same units used in the accumulation term) that enter and leave the system over a period of time Δt. The higher the temperature of the stream fed, the greater the amount of heat that enters the system. Likewise, the higher the mass flow rate entering the tank, the greater the amount of heat fed into the tank. Another parameter that affects the heat flow is the characteristic of the fluid, which may be given by its specific heat. So the amount of energy that enters the tank in a period of time Δt can be given by:

$$E = \underbrace{2\left(\frac{\text{m}^3}{\text{h}}\right)\rho\left(\frac{\text{kg}}{\text{m}^3}\right)}_{\text{mass flow rate}} c_p \left(\frac{\text{J}}{\text{kg }^\circ\text{C}}\right) 50\,(^\circ\text{C})\,\Delta t(\text{h})$$

$$E = 100\rho c_p \Delta t \quad (\text{J})$$

Analogously, one can obtain the amount of energy that leaves the system. We do not know the temperature of the water that leaves the system (which equals the temperature inside the tank, because it is perfectly mixed), so we call this generic temperature T. Note that we have also considered this temperature T in the accumulation term.

$$L = 2 \left(\frac{m^3}{h}\right) \underbrace{\rho \left(\frac{kg}{m^3}\right)}_{\text{mass flow rate}} c_p \left(\frac{J}{kg\ ^{\circ}C}\right) T(^{\circ}C)\ \Delta t(h)$$

$$L = 2\rho c_p T \Delta t \quad (J)$$

It is well known that ρ and c_p depend on the temperature; however, in our example we will assume that the ρ and c_p of the water fed into the tank are equal to the ρ and c_p of the water that leaves the tank.

The application of the conservation law generates the following energy balance (in joules):

$$V\rho c_p \frac{dT}{dt} \Delta t = 100\rho c_p \Delta t - 2\rho c_p T \Delta t \tag{3.2}$$

Simplifying and rearranging it yields:

$$\frac{dT}{dt} = \frac{100 - 2\,T}{V}$$

In our case the volume of water inside the tank stays constant and is equal to 10 m^3, so the ODE becomes:

$$\frac{dT}{dt} = 10 - 0.2\,T$$

This ODE can be solved analytically or numerically using the initial condition (at $t = 0$, $T = 20\ ^{\circ}C$) to generate the temperature profile shown in Fig. 3.5, which represents the water temperature inside the tank over time.

In Fig. 3.5, one can observe that after around 25 h, the system reaches a steady state because the temperature does not change with time anymore.

If we want to know the temperature of the liquid inside the tank in a steady state without plotting the curve, the conservation law for a steady state has to be used ($E - L = 0$). The energy balance is developed without considering the accumulation term (no variation of temperature with time) to yield (compare this with Eq. 3.2):

$$100\rho c_p - 2\rho c_p T = 0 \quad\quad \text{or}$$

$$100 - 2\,T = 0$$

So the temperature of the liquid inside the tank after reaching a steady state is 50 $^{\circ}C$, as predicted by inspection and observed in Fig. 3.5.

If the volume, the concentration, and the temperature change simultaneously, three ODEs have to be solved simultaneously in order to predict the system's behavior. Imagine the problem presented in Example 3.2, but consider that initially the temperature of the water in the tank is 20 $^{\circ}C$ and the temperature of the fluid fed

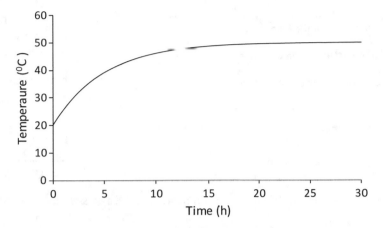

Fig. 3.5 Temperature profile of water inside the tank over time

into the tank is 50 °C. The volume and concentration equations are the same as those obtained in Example 3.2, and the energy balance has to be solved considering that the volume is not constant and changes over time. Assuming that the density and the specific heat do not change significantly with the concentration of NaOH and temperature, one can obtain the following set of equations to represent the system:

$$\frac{dV}{dt} = 1$$

$$\frac{dx}{dt} = \frac{0.06 - 3x}{V}$$

$$\frac{dT}{dt} = \frac{150 - 3T}{V}$$

with the initial conditions: At $t = 0$, $V = 10$ m^3, $x = 0$ kg/m^3, and $T = 20$ °C.

The concentration and temperature equations do not depend on each other, but both depend on the volume balance.

After understanding these introductory examples, you are ready to revisit some concepts needed to model more complex problems. Sections 3.2 and 3.3 will deal with convective heat transfer and chemical reactions, respectively.

3.2   Some Concepts About Convective Heat Exchange

All examples presented up to this point have considered an adiabatic system, i.e., insulated tanks with no heat exchange with the environment. In real chemical plants, heat exchange between the system and the environment is very common.

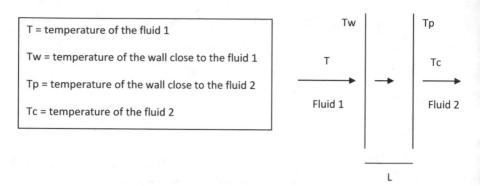

Fig. 3.6 Flow of energy (shown by the *arrows*) from the fluid at $T\,°C$ to the fluid at $Tc\,°C$

In order to promote the addition or removal of energy, jacketed vessels are often used. These vessels are tanks designed to control the temperature of their contents. When there is heat exchange with the environment or with a jacket, this flow of energy has to be considered in the energy balance.

In order to better understand the heat transfer between two fluids at different temperatures, observe Fig. 3.6, which shows a wall with a certain thickness L separating two fluids with temperatures T and Tc °C. Assume that T is greater than Tc, so there will be a flow of energy from the left side to the right side, as depicted by the arrows.

The flow of energy by conduction, which will be further detailed in Chap. 4, deals with the transfer of energy from molecule to molecule in materials in which molecules have little or nearly no mobility. This is the case with solid materials. Observing Fig. 3.6 one can conclude that the energy flows through the wall by conduction, and that there is a profile of temperature along the thickness of the wall ($Tw > Tp$, being Tw and Tp temperatures at both surfaces of the wall). Both fluids have molecules with more mobility, so the flow of energy in these two media occurs mostly by convection.

Heat flow by conduction or convection is directly proportional to the driving force and inversely proportional to the resistance. In Fig. 3.6, the driving force for the flow of energy through fluid 1 is $T - Tw$. Likewise, the driving forces for the energy flow through the wall and through fluid 2 are $Tw - Tp$ and $Tp - Tc$, respectively. Fluids 1 and 2 and the wall will offer resistance to the flow of energy.

Along the wall, the flow of energy will depend on the properties of the wall, like thermal conductivity. Materials with higher thermal conductivity offer low resistance to heat flow by conduction (it is well known that insulating materials such as Styrofoam present low thermal conductivity). Moreover, the longer the distance that the heat has to go through by conduction, the greater the resistance to the flow of energy. In this way, the resistance to the flow of energy by conduction can be represented in this example by L/k (this expression is valid for Cartesian coordinates), in which L is the thickness of the wall and k is the thermal conductivity of the wall.

The flow of energy by convection through fluids 1 and 2 will depend not only on the properties of the fluids (such as viscosity, specific heat, and density) but also on the operating conditions (for example, the higher the fluid velocity, the better the heat exchange). Analogous to k for the conduction, the parameter that indicates if the fluid is effective for energy transportation by convection is the *heat transfer coefficient* (h), which is a function of the properties of the fluids and operating conditions. The higher the value of h is, the more effective the heat flow by convection is. In this way, the resistance for the convective flow is $1/h$. There are many different correlations to obtain the heat transfer coefficient (h); however, in this book the values of h will always be informed.

Having in mind the concepts presented above, one can obtain the three energy flows shown in Fig. 3.6 (α means proportional).

$$\text{Flow}_{\text{fluid 1}} \alpha \; \frac{T - Tw}{\frac{1}{h_{\text{fluid1}}}}; \qquad \text{Flow}_{\text{wall}} \alpha \; \frac{Tw - Tp}{\frac{L}{k}}; \qquad \text{Flow}_{\text{fluid 2}} \alpha \; \frac{Tp - Tc}{\frac{1}{h_{\text{fluid2}}}}$$

The total resistance to the heat flow is given by:

$$R = \frac{1}{h_{\text{fluid1}}} + \frac{L}{k} + \frac{1}{h_{\text{fluid2}}}$$

in which (using international units):

h_{fluid1} = heat transfer coefficient for fluid 1 $\left(\dfrac{J}{s\; m^2\; {}^\circ C}\right)$

h_{fluid2} = heat transfer coefficient for fluid 2 $\left(\dfrac{J}{s\; m^2\; {}^\circ C}\right)$

k = thermal conductivity $\left(\dfrac{J}{s\; m\; {}^\circ C}\right)$

L = thickness of the wall (m)

Usually the resistance to conduction is irrelevant if compared with the resistance to convection in chemical plants, because pieces of equipment in an industry are usually built with material with high thermal conductivity (usually metals) and have thin walls, so the total resistance becomes:

$$R = \frac{1}{h_{\text{global}}} = \frac{1}{h_{\text{fluid1}}} + \frac{1}{h_{\text{fluid2}}}$$

h_{global} is called the global heat transfer coefficient, and it is also represented by U or h.

So the flow of energy from fluid 1 to fluid 2 is proportional to the total driving force ($T - Tc$) and inversely proportional to the total resistance (R).

$$\text{Flow}_{\text{total}} \alpha \; \frac{T - Tc}{\frac{1}{U}}$$

The factor of proportionality is the area (A) through which there is heat exchange, and this area is perpendicular to the direction of the heat flow. Therefore, the convective heat flow (called Q below) is given by:

$$Q = UA\,(T - Tc) = UA\,\Delta T$$

in which (using international units):

Q = convective heat flow (J/s)
U = global heat transfer coefficient (J/s m^2 °C)
A = area of convective heat transfer (m^2)
ΔT = difference in temperature (°C)

It is important to observe that, as the heat flow is continuous, we can say:

Heat flow in fluid 1 = Heat flow in the wall = Heat flow in fluid 2 = Heat flow

Now we are ready to again solve Example 3.3, but this time considering a noninsulated system with loss of heat to the environment.

It is important to point out that the concepts presented in Sect. 3.2 are the minimum necessary to develop mathematical models of systems that present heat exchange by convection. The reader can find specific and detailed literature on this subject elsewhere (Kern 1950; Incropera et al. 2006; Bird et al. 2007; Welty et al. 2007).

Example 3.4 Energy Balance in a Noninsulated Stirred Tank with Convective Heat Transfer

The next example will give another step forward in complexity by considering a noninsulated tank. Let us consider Example 3.3, but this time assuming that the tank exchanges heat with the environment, which is at a temperature of 15 °C (Fig. 3.7).

One wants to know how the water temperature inside the tank changes over time and what the temperature is in a steady state.

Assume that the global heat transfer coefficient (U) between the liquid inside the tank and the environment is equal to 30 (J/s m^2 °C). Consider constant values for the density ($\rho = 1000$ kg/m^3) and specific heat ($c_p = 4184$ J/kg °C) of the fluid inside the tank, even with changes in temperature. Assume that the heat exchange area (A)

Fig. 3.7 Stirred tank exchanging heat with the environment (noninsulated tank)

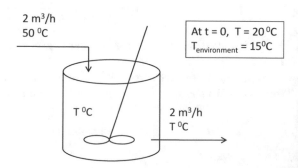

with the environment is equal to $40\,\mathrm{m}^2$ and the volume of liquid inside the tank (V) is equal to $V = 10\,\mathrm{m}^3$.

Solution:

In this problem, we need to again solve Example 3.3, but this time considering also the convective heat transfer with the environment, which is at 15 °C.

The application of the conservation law to our problem keeps yielding $E - L = A$, but this time, besides the terms considered in Example 3.3, we have to consider the term related to the heat exchange with the environment.

The accumulation term is calculated in the same way as was done before, and yields:

$$A = V\rho c_p \frac{dT}{dt} \Delta t \quad (\mathrm{J})$$

The amount of energy that enters the tank is the same as that developed in Example 3.3:

$$E = 2\left(\frac{\mathrm{m}^3}{\mathrm{h}}\right)\rho\left(\frac{\mathrm{kg}}{\mathrm{m}^3}\right)c_p\left(\frac{\mathrm{J}}{\mathrm{kg}\,°\mathrm{C}}\right)50\,(°\mathrm{C})\,\Delta t(\mathrm{h})$$

$$E = 100\rho\,c_p\Delta t \quad (\mathrm{J})$$

The amount of energy that leaves the tank due to the fluid leaving the tank, obtained in Example 3.3, has to be considered in this example too:

$$L = 2\left(\frac{\mathrm{m}^3}{\mathrm{h}}\right)\rho\left(\frac{\mathrm{kg}}{\mathrm{m}^3}\right)c_p\left(\frac{\mathrm{J}}{\mathrm{kg}\,°\mathrm{C}}\right)T(°\mathrm{C})\,\Delta t(\mathrm{h})$$

$$L = 2\rho c_p T \Delta t \quad (\mathrm{J})$$

However, the flow of energy that leaves the tank by convection in a period of time Δt has to be considered also, and it is given by:

$$L = U\left(\frac{\mathrm{J}}{\mathrm{h}\,\mathrm{m}^2\,°\mathrm{C}}\right)A\,(\mathrm{m}^2)\,(T - 15)\,(°\mathrm{C})\,\Delta t(\mathrm{h})$$

$$L = U A\,(T - 15)\Delta t \quad (\mathrm{J})$$

So the application of the conservation law to this case yields:

$$\underbrace{V\rho c_p \frac{dT}{dt}\Delta t}_{\text{Accumulates (A)}} = \underbrace{100\rho c_p\Delta t}_{\text{Enters (E)}} - \underbrace{2\rho c_p T \Delta t - U A(T - 15)\Delta t}_{\text{Leaves (L)}} \qquad (\mathrm{J}) \qquad (3.3)$$

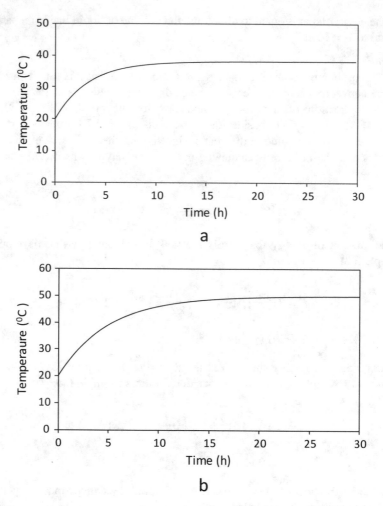

Fig. 3.8 Profiles of temperature over time: (**a**) for a stirred tank exchanging heat with the environment; (**b**) for the insulated tank presented in Example 3.3

Rearranging:

$$\frac{dT}{dt} = \frac{1}{V}\left[100 - 2T - \frac{UA}{\rho c_p}(T - 15)\right]$$

The initial condition to solve this ODE remains the same: at $t = 0$, $T = 20\,°C$.

After solving this ODE, the temperature profile over time can be seen in Fig. 3.8a. Observe that the system reaches a steady state at around 40 °C, which is a lower temperature than the one in Example 3.3, where there was no heat exchange with the environment. Figure 3.8b again shows Fig. 3.5, to simplify the comparison.

Now a *simple convention* to help the development of a model will be introduced. Imagine that the environment temperature is 25 °C, instead of 15 °C. In this case, since the initial temperature of the water is 20 °C, maybe in the beginning the heat flows from the environment to the tank. However, after some time, as the tank is fed with a fluid at 50 °C, the heat flows from the tank to the environment. Should we consider this heat flow by convection entering (E; plus sign) or leaving (L; minus sign) the tank? In order to deal with situations like that, a simple convention can be used: always add (plus sign) the convective terms in the energy balance, but always consider the difference in temperature as the environment (surrounding) tempera-ture minus the system temperature, in this order, as represented below:

$$U A \left(T_{env} - T \right)$$

in which T_{env} is the temperature of the environment or the jacket (or surrounding) and T is the temperature of the control volume. So if $T_{env} > T$, the convection term is positive and heat is added to the system. If $T_{env} < T$, the convection term is negative and heat is removed from the system. If at some point $T_{env} = T$, there is no heat exchange between the tank and the environment.

Using this convention, Eq. (3.3) can be rewritten as:

$$\underbrace{V\rho c_p \frac{dT}{dt} \Delta t}_{\text{Accumulates (A)}} = \underbrace{100\rho c_p \Delta t}_{\text{Enters (E)}} - \underbrace{2\rho c_p T \Delta t}_{\text{Leaves (L)}} + \underbrace{UA(15 - T)\Delta t}_{\text{Heat by Convection}} \quad \text{(J)} \qquad (3.4)$$

If we want to model this system to obtain the temperature of the tank when a steady state is reached, the energy balance has to be calculated without considering the accumulation term, and the following equation is obtained (compare this with Eq. 3.4):

$$100\rho c_p - 2\rho c_p T + UA(15 - T) = 0 \quad \text{(J/h)}$$

Considering the numerical values for the parameters and solving this algebraic equation, we can obtain that the temperature in a steady state is 38.08 °C, as previously observed in Fig. 3.8a.

Now let us explore the behavior of this tank a little further by studying another possible situation, suggested in Example 3.5.

Example 3.5 Energy Balance Considering Convective Heat Transfer and No Inlet or Outlet Flow Rates
This example revisits Example 3.4 and assumes that when the water temperature inside the tank reaches 38.08 °C (the temperature in a steady state), the input and output valves are closed. If this is the case, how does the water temperature decrease over time? Assume again that the tank is perfectly mixed and exchanges heat with the environment, which is at 15 °C.

Solution:
The energy balance presented in Eq. (3.4) is simplified, removing the terms of inflow and outflow of the water, to yield:

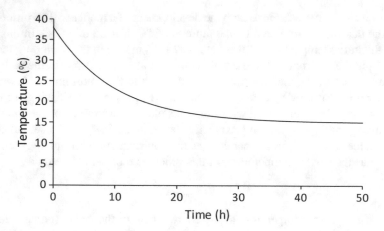

Fig. 3.9 Profile of water temperature over time

$$\underbrace{V\rho c_p \frac{dT}{dt}\Delta t}_{\text{Accumulates (A)}} = \underbrace{UA(15-T)\Delta t}_{\text{Heat by Convection}} \text{(J)} \tag{3.5}$$

Equation (3.5) can be rewritten as:

$$\frac{dT}{dt} = \frac{UA}{V\rho c_p}(15-T) \tag{3.6}$$

Solving Eq. (3.6), considering the initial condition (at $t = 0$, $T = 38.08\,°C$), we can find that it takes around 50 h until the water reaches the environment temperature (15 °C), as shown Fig. 3.9. In order to solve Eq. (3.6), the same numerical values used in Example 3.4 are adopted ($U = 30\,J/s\,m^2\,°C$, $c_p = 4184\,J/kg\,°C$, $\rho = 1000\,kg/m^3$, $A = 40\,m^2$, $V = 10\,m^3$).

3.3 Some Concepts About Chemical Kinetics and Reactors

All systems studied so far have not considered chemical reaction. When there is a chemical reaction, the temperature of the system can change, because the reactions can be exothermic or endothermic. Besides, chemical reactions cause changes in the concentrations of reactants and products, although the total mass of the system remains constant.

It is not the objective of this book to explore the kinetics and reactions issue in detail. Herein a very few concepts, in a very simplified way, will be presented, just to allow us to develop mathematical models for systems with chemical reactions. The reader can find very interesting books in the literature dealing with chemical reactions and kinetics, such as Froment and Bischoff (1990), Fogler (1999), Davis and Davis (2003), and Hill and Root (2014), just to mention a few.

3.3.1   Some Concepts About Kinetics of Chemical Reactions

Imagine an irreversible chemical reaction where a reagent A is transformed into a product B, as shown below:

$$A \xrightarrow{k} B \tag{3.7}$$

in which k is the kinetic rate constant.

In the above reaction, one mol of reactant A produces one mol of product B (a stoichiometric reaction). The speed at which the reaction occurs is given by the *rate of chemical reaction* (r) and represents the number of moles consumed per volume per time. The rate of irreversible reactions is given by the kinetic rate constant (k) multiplied by the concentrations of the reagents, considering the stoichiometry of the reaction.

In a first-order reaction, in which there is only one kind of reactant and the stoichiometric coefficient of the reactant is 1 (as occurs in Eq. 3.7), the constant rate (k) has the dimension of time^{-1}. Using international units, the reaction rate can be expressed by:

$$\boxed{\text{Reaction rate} = r = k\, C_A^{\nu_A} = k\, C_A}$$

C_A = reactant concentration (mol/m^3)

k = rate constant $(1/\text{s})$

r = reaction rate $(\text{mol/m}^3\ \text{s})$

ν_A = partial order of reaction (dimensionless)

In this book we will consider the partial orders of reaction equal to the stoichiometric coefficients of the reactant (in our case, $\nu_A = 1$), but sometimes they depend on the reaction mechanism and can be determined experimentally.

Other examples of irreversible first-order reactions are shown below:

$$A \xrightarrow{k} 2B \qquad \Rightarrow \quad r = k\, C_A \tag{3.8}$$

$$A \xrightarrow{k} B + C \quad \Rightarrow \quad r = k\, C_A \tag{3.9}$$

Observe in Eq. (3.8) that reactant A is consumed at a rate $k\, C_A$ $(\text{mol/m}^3\ \text{s})$ to produce B at a rate $2\, k\, C_A$ $(\text{mol/m}^3\ \text{s})$, because one mol of A generates 2 mols of B, according to the stoichiometry of Eq. (3.8).

The unit of the rate constant (k) depends on the order of the reaction. For example, if the irreversible reaction is of second order (two mols of different or equal reactants are needed to produce the product), the rate constant has the units $\text{m}^3/\text{mol s}$ (see Eq. 3.10).

$$A + B \xrightarrow{k} C \tag{3.10}$$

Reaction rate $= r = kC_A^{\nu_A}C_B^{\nu_B} = kC_A C_B$

C_A = concentration of reactant A (mol/m^3)
C_B = concentration of reactant B (mol/m^3)
k = rate constant (m^3/mol s)
r = reaction rate (mol/m^{3}s)
$\nu_A = 1$ and $\nu_B = 1$ (dimensionless)

Other examples of irreversible second-order reactions are shown below:

$$2A \xrightarrow{k} B \qquad \Rightarrow \quad r = k\,C_A^2$$

$$A + B \xrightarrow{k} C + D + E \quad \Rightarrow \quad r = k\,C_A C_B$$

The rate constant k usually follows the Arrhenius law, varying exponentially with temperature:

$$k = k_0 \exp\left(-\frac{E_A}{RT}\right)$$

k = rate constant
k_0 = pre−exponential factor
E_A = activation energy
R = gas constant
T = absolute temperature (K)

Note that the reaction rate (r) depends not only on the concentration of the reactants but also on the temperature. Thus, when a non isothermal chemical reactor is modeled, the mass and energy balances must be solved simultaneously.

The chemical reaction can also be reversible or can present a mechanism composed of many steps. More complex kinds of kinetics are not explored in this book but can be found in any book regarding kinetics and reactors.

3.3.2 Some Concepts About Chemical Reactors

There are a lot of types of chemical reactors, but this book will cover only the most common types and their operation modes.

Imagine a cylindrical stirred tank reactor. Ideally, it can be considered that the agitation of this reactor is perfect (a lumped-parameter system), which gives us the first class of reactors: stirred tank reactors (STRs). Basically, these stirred tank reactors can operate in three different ways:

- *Batch*: when all reactants are added to the reactor at once before the beginning of the reaction, and there is no addition of reactants or withdrawal of products as the reaction occurs (see Scheme A).
- *Continuous*: when the reactants and products are continuously fed and withdrawn. This is the system known as a continuous stirred tank reactor (CSTR) (see Scheme B).
- *Semi-batch or fed-batch*: when there is no removal of products during the reaction, but the reagents can be added as the reaction proceeds (see Scheme C).

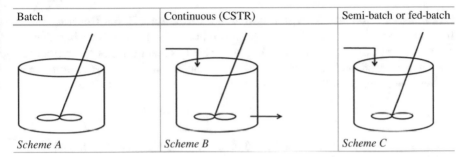

Batch	Continuous (CSTR)	Semi-batch or fed-batch
Scheme A	*Scheme B*	*Scheme C*

There is another type of reactor widely used in industries: the tubular reactor. As the name says, this reactor has the shape of a tube and its dimensions (length and diameter) depend on the type of process or product. A simplified scheme for a tubular reactor can be seen in Fig. 3.10, which shows reactants entering through one side of the reactor and products (and no consumed reagents) leaving through the opposite side. The reaction proceeds as the reaction mixture moves forward inside the reactor.

Once again, it is important to point out that the information about chemical reactors presented in this book is only the minimum necessary to enable the development of mathematical models for systems that present chemical reactions. The reader can find very interesting books on chemical reactors and kinetics in the specific literature (Levenspiel 1999; Froment and Bischoff 1990; Missen et al. 1999; Davis and Davis 2003, etc.).

The last three examples in this chapter will present problems of lumped parameters in which chemical reactions occur.

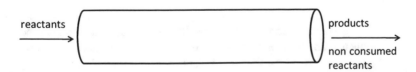

Fig. 3.10 Scheme of a tubular reactor

Fig. 3.11 Batch stirred
tank reactor

$$t = 0: \quad C_A = C_{A0} \ (\text{mol/m}^3)$$
$$C_B = C_C = 0$$
$$T = T_0 \ (^0\text{C})$$

Example 3.6 Mass and Energy Balance in an Adiabatic Batch Reactor

Imagine a batch stirred tank reactor, as shown in Fig. 3.11. Let us consider that the reactor is perfectly mixed and that there is no heat exchange with the environment (an adiabatic system). This reactor is used to produce component B, according to an irreversible chemical reaction:

$$A \xrightarrow{k} 2B + C$$

in which A is the reactant, B is the desired product, and C is an undesirable subproduct, which must be removed using a separation system to be installed after the reactor.

The chemical reaction is exothermic and follows the Arrhenius law. In the beginning, there are no products in the reactor, and the initial temperature and concentration of reactant A are T_0 ($^\circ$C) and C_{A0} (mol/m^3), respectively. Let us assume that the volume of the reactional mixture is equal to V (m^3). For simplicity, let us consider that the densities of compounds A, B, and C are practically the same and constant with temperature, so it is possible to assume that the volume of liquid inside the reactor remains the same.

Develop a mathematical model to represent this reactor in order to obtain the profiles of C_A, C_B, C_C, and T over time.

Solution:

Since the reactor is perfectly mixed, its content is homogeneous and the control volume is the entire tank (a lumped-parameter problem).

Let us start by calculating the mass balance. The mass balance for reactant A and products B and C can be obtained by applying the conservation law equation (E − L + G − C = A).

As we are modeling a batch reactor, in which there is no entry or exit of A, B, and C (observe Fig. 3.11), the terms E (entry) and L (exit) in the three mass balances are equal to zero. In addition, for the mass balance of reactant A, we must consider the consumption term, but a generation term does not exist (an irreversible reaction). On the other hand, for products B and C an opposite situation occurs (there is no consumption term, but a generation term exists). So the mass balance for A, B, and C becomes:

Mass Balance for reactant A: Accumulation of A = −Consumption of A
Mass Balance for product B: Accumulation of B = Generation of B
Mass Balance for product C: Accumulation of C = Generation of C

Let us start calculating the accumulation terms in a period Δt (s), as follows:

For reactant A:

t	$t + \Delta t$	Accumulation of A in Δt (mol)
VC_A	$VC_A + \dfrac{d(VC_A)}{dt}\Delta t$	$V\dfrac{d(C_A)}{dt}\Delta t$

For product B:

t	$t + \Delta t$	Accumulation of B in Δt (mol)
VC_B	$VC_B + \dfrac{d(VC_B)}{dt}\Delta t$	$V\dfrac{d(C_B)}{dt}\Delta t$

For product C:

t	$t + \Delta t$	Accumulation of C in Δt (mol)
VC_C	$VC_C + \dfrac{d(VC_C)}{dt}\Delta t$	$V\dfrac{d(C_C)}{dt}\Delta t$

Now we need to obtain the consumption and generation terms.

The chemical reaction is of first order so the rate constant has the dimension of time^{-1}. Using international units, the reaction rate is given by:

$$r = k\, C_A^{\nu_A}$$

C_A = concentration of reactant A (mol/m^3)

k = rate constant (s^{-1})

ν_A — stoichiometry coefficient = 1 (dimensionless)

r = reaction rate (mol/m^3 s)

Since for each mol consumed of reactant A, two mols of product B and one mol of product C are generated, the rate at which reactant A is consumed (r_A) is equal to the rate at which product C is generated (r_C), but the rate at which B is produced (r_B) is twice the rate of consumption of reactant A. So:

$$r_A = r_C = k\, C_A$$
$$r_B = 2\, r_A = 2\, k\, C_A$$

Observe that the reaction rate is given in the units mol/m^{3}s, and the accumulation terms is given in the units mol, so before using these two terms in the mass balance, we must convert them to the same units.

Observe that we calculated the amounts of A, B, and C accumulated in the entire volume of the reactor in a period of time Δt. The reaction rate gives us the amounts of A, B, and C consumed or produced per m^3 of reactor per second. So, to be consistent, we have to multiply the reaction rate by the volume of the reactor and by the period of time Δt:

Amount of A consumed in the reactor in a time Δt: $kC_A V \Delta t$ (mol)

Amount of B produced in the reactor in a time Δt: $2kC_A V \Delta t$ (mol)

Amount of C produced in the reactor in a time Δt: $kC_A V \Delta t$ (mol)

So the material balances for reactant A and products B and C become:

Balance of A (mol): Accumulation of A = −Consumption of A:

$$V\frac{d(C_A)}{dt}\Delta t = -kC_A V \Delta t$$

Balance of B (mol): Accumulation of B = Generation of B:

$$V\frac{d(C_B)}{dt}\Delta t = 2kC_A V \Delta t$$

Balance of C (mol): Accumulation of C = Generation of C:

$$V\frac{d(C_C)}{dt}\Delta t = kC_A V \Delta t$$

Simplifying terms, this yields:

Balance of A (mol):	$\dfrac{dC_A}{dt} = -k\,C_A$
Balance of B (mol):	$\dfrac{dC_B}{dt} = 2\,k\,C_A$
Balance of C (mol):	$\dfrac{dC_C}{dt} = k\,C_A$

Observe that the balances of products B and C depend on the concentration of reactant A, so they must be solved simultaneously with the balance of A.

This system of ODEs has to be solved analytically or numerically in order to obtain profiles of the concentrations of A, B, and C over time. To do that, the following initial conditions have to be used:

$$\text{At } t = 0, C_A = C_{A_0}\left(\text{mol/m}^3\right), C_B = C_C = 0$$

Note, however, that all mass balance equations show the term k (the rate constant), which follows the Arrhenius law and therefore varies exponentially with temperature. Thus, in order to properly obtain the mass balance, the energy balance must be solved simultaneously (this is not necessary only for isothermal systems).

For the energy balance, the conservation law equation (E − L + G − C = A) also has to be applied considering the entire reactor as the control volume. As we are modeling a batch reactor, there is no heat being added or withdrawn by input and output streams (see Fig. 3.11), so this kind of heat cannot be considered in the terms

E (entry) and L (exit) in the conservation law equation. Besides, as the system is adiabatic, the reactor does not exchange heat with the environment, so heat exchange by convection does not exist. In this way, the energy balance will consider the amount of energy accumulated in the system and the heat generated or consumed by the chemical reaction. If the reaction is exothermic, heat will be added to the system, but if the reaction is endothermic, heat will be removed from the system.

As was done for the material balance, let us start by calculating the energy accumulation in the reactor over a period of time Δt (s) (observe that the accumulation term is the same as that obtained in Examples 3.3, 3.4, and 3.5):

Accumulation of energy:

t (s)	$t + \Delta t$ (s)	Accumulation of energy in Δt (J)
$V \rho c_p T$	$V \rho c_p T + \dfrac{d(V \rho c_p T)}{dt} \Delta t$	$\dfrac{d(V \rho c_p T)}{dt} \Delta t$

$V =$ volume of the liquid inside the reactor (m^3)
$\rho =$ density of the liquid inside the reactor (kg/m^3)
$c_p =$ specific heat of the fluid inside the reactor (J/kg °C)
$T =$ temperature of the liquid inside the reactor (°C)

Observe that the accumulated heat is given in joules (see units of V, ρ, c_p, and T). The c_p value of the fluid depends on the temperature and on the composition of the liquid inside the tank, but, for simplicity, let us assume that c_p does not vary significantly with temperature, and that $c_{pA} \cong c_{pB} \cong c_{pC}$, so we can assume that c_p of the liquid is constant throughout the reaction. As mentioned earlier, it is assumed that V and ρ remain constant, so the energy accumulation term becomes:

$$V \rho c_p \frac{dT}{dt} \Delta t \quad (J)$$

Now let us calculate the term related to the heat liberated or absorbed due to the chemical reaction.

As said earlier, the rate at which the reaction occurs is given by the reaction rate r. The reaction rate in our case is kC_A, and its unit is mol/m^3 s.

The heat liberated or absorbed in a chemical reaction depends on the *enthalpy of the reaction*, also known as the *energy change of the reaction*, $(\Delta H)_R$, which is the difference between the total enthalpy of the products and the total enthalpy of the reactants. Usually, the enthalpy of the reaction is given in units of energy per mol (for example, J/mol) and means the energy liberated or absorbed by each mol reacted. In this way, the enthalpy of the reaction (J/mol) has to be multiplied by the reaction rate (mol/m^3 s) in order to obtain the total heat generated or absorbed in a chemical reaction (J/m^3 s), as shown below.

Reaction rate: $r = kC_A$ (mol/m^3 s)
Enthalpy of the reaction: $(\Delta H)_R$ (J/mol)
Heat liberated or absorbed in a chemical reaction: $kC_A(\Delta H)_R$ (J/m^3 s)

For the batch reactor that is being modeled, the amount of heat liberated or absorbed by the chemical reaction over a period of time Δt (s) in the control volume (the entire reactor) is:

Heat liberated or absorbed: $k\,C_A(\Delta H)_R V\,\Delta t$ (J)

Before combining the accumulation and heat of reaction terms in the conservation equation, let us create a *sign convention*. It is well known that $(\Delta H)_R$ is negative for exothermic reactions and positive for endothermic reactions. In this way, we will consider a minus sign in front of $(\Delta H)_R$ and we will always add the liberated or absorbed heat term in the conservation equation. In this way, the energy balance becomes:

$$V\rho c_p \frac{dT}{dt}\Delta t = V k C_A(-\Delta H)_R \Delta t \qquad (3.11)$$

In our case the chemical reaction is exothermic $((\Delta H)_R < 0)$. As we have already considered the minus sign in $(\Delta H)_R$, the term of the heat of the reaction will become positive (heat being added to the system). If the reaction were endothermic $((\Delta H)_R > 0)$, the term of the heat of the reaction would be negative (heat being removed from the system), because of the minus sign considered in the convention. Using this convention, we do not need to worry about defining whether the heat is being generated or consumed, and the model becomes generic.

So, from now on, the conservation law for the energy balance will be written as:

$$\mathrm{E - L + G/C = A}$$

The term G/C represents the amount of energy generated or absorbed, and it will always be added to the conservation law.

Simplifying and rearranging the terms of Eq. (3.11), the energy balance becomes:

$$\frac{dT}{dt} = \frac{kC_A(-\Delta H_R)}{\rho c_p}$$

In order to solve this equation, the initial condition to be used is: at $t=0$, $T = T_0$ (°C).

The ODEs system that represents this reactor is:

Balance of A (mol): $\dfrac{dC_A}{dt} = -kC_A$

Balance of B (mol): $\dfrac{dC_B}{dt} = 2kC_A$

Balance of C (mol): $\dfrac{dC_C}{dt} = kC_A$

Balance of Energy (J): $\dfrac{dT}{dt} = \dfrac{kC_A(-\Delta H_R)}{\rho c_p}$

Initial conditions: at $t = 0$: $C_A = C_{A_0}$, $C_B = C_C = 0$, $T = T_0$

Observe that the energy balance depends on the concentration of the reactant A, and, as said earlier, k depends on the temperature, so the mass and energy balances must be solved simultaneously.

If we assume numerical values for all parameters in the model (let us suppose $E_A = 48{,}500$ J/mol, $R = 8.314$ J/mol K, $k_0 = 8.20 \times 10^7$ min^{-1}, $\Delta H_R = -72{,}800$ J/mol, $c_p = 1750$ J/kg K, $\rho = 880$ kg/m^3) and for the initial conditions ($T_0 = 300$ K, $C_{A_0} = 100$ mol/m^3, and $C_{B_0} = C_{C_0} = 0$), the system of four equations can be solved and profiles of the concentration and temperature over time can be obtained (see Fig. 3.12).

Observe that after 15 min, all of reactant A is consumed ($C_A = 0$), so the concentrations of B and C and the temperature do not change anymore, because there is no chemical reaction. Note also that the concentration of B is twice the concentration of C, as expected due to the stoichiometry of the reaction.

The next example deals with a CSTR equipped with a cooling jacket to control the temperature of the reactor.

Example 3.7 Mass and Energy Balance in a CSTR with a Cooling Jacket Operating in a Steady-State Regime

Imagine a CSTR operating in a steady state in which the exothermic reaction $A + B \xrightarrow{k} C$ takes place. The reactor has a cooling jacket to control its temperature (see Fig. 3.13). A solution with reactants A and B is fed into the reactor with a flow rate Q (m^3/min) and temperature T_{in} (K) (at T_{in}, reactants A and B are not able to react). The concentrations of reactants A and B in this feed solution are $C_{A_{in}}$ and $C_{B_{in}}$ (mol/m^3), respectively. The fluid leaves the reactor at the same flow rate (Q) and contains product C as well as the A and B that may be not totally consumed. Assume that the density and specific heat for all compounds are almost the same and do not vary as the reaction occurs ($\rho_A = \rho_B = \rho_C = \rho$ and $c_{pA} = c_{pB} = c_{pC} = c_p$). The volume of the liquid inside the reactor is V (m^3) and does not change over time. The cooling fluid is fed into the jacket at a flow rate Q_J (m^3/min) and at a temperature Tj_{in} (K). Assume that the density and specific heat of the cooling fluid (ρ_j and c_{pj}) do not vary during the entire process. The global heat transfer coefficient between the cooling fluid and the reaction mixture is U (J/min m^2 K). The area from where the heat exchange occurs is A (m^2). Find the system of equations that represents this reactor in a steady state, i.e., find the mass balance for A, B, and C, and the energy balance for the reactor and for the jacket.

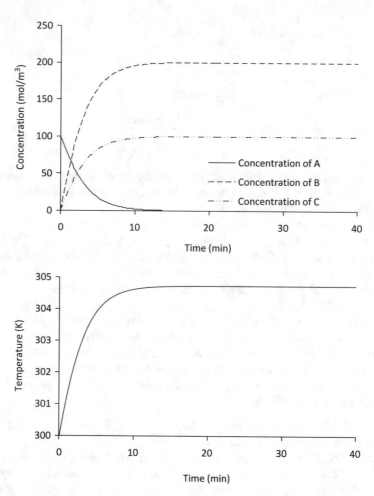

Fig. 3.12 Batch reactor behavior: (**a**) profiles of concentrations of A, B, and C over time; (**b**) profile of temperature over time

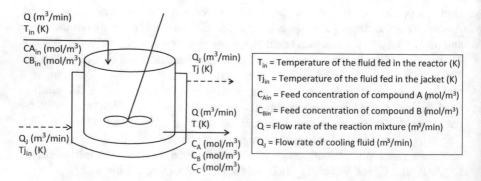

Fig. 3.13 Continuous stirred tank reactor (CSTR) with a cooling jacket

Solution:

Let us start with the mass balance for A, B, and C. As the CSTR operates in a steady state, there is no variation in the concentration over time, and the accumulation terms for the three compounds are zero. So the conservation law equation applied to the control volume (the entire reactor) yields:

$$\text{Mass Balance for A:} \qquad E - L - C = 0$$
$$\text{Mass Balance for B:} \qquad E - L - C = 0$$
$$\text{Mass Balance for C:} \qquad E - L + G = 0$$

The amounts of A, B, and C that enter (E) and leave (L) the reactor can be obtained by multiplying the volumetric flow rate (m^3/min) by the concentrations of A, B, or C (mol/m^3) to yield the number of moles of each compound that enter or leave the reactor per minute (mol/min).

Compound	Enters (mol/min)	Leaves (mol/min)
A	QC_{Ain}	QC_A
B	QC_{Bin}	QC_B
C	QC_{Cin}	QC_C

Remember that compound C is not fed into the reactor, so $C_{Cin} = 0$; however, we will keep C_{Cin} in the mass balances in order to obtain a generic model.

The generation and consumption terms are obtained from the reaction rate. As there are two reactants, the reaction rate is given by kC_AC_B (mol/m^3 min). Observe that this chemical reaction is of second order, so the unit for the rate constant (k) is m^3/mol min. The reaction rate must be multiplied by the control volume (the volume of the entire reactor) to obtain the generation and consumption terms in mol/min (the same units used for the terms E and L). The following table shows the reaction terms.

Compound	Consumption (mol/min)	Generation (mol/min)
A	$k\,C_A\,C_B\,V$	0
B	$k\,C_A\,C_B\,V$	0
C	0	$k\,C_A\,C_B\,V$

So the mass balance for A, B, and C becomes:

$$\text{Mass balance for A (mol/min):} \qquad Q(C_{Ain} - C_A) - kC_AC_BV = 0 \qquad (3.12)$$
$$\text{Mass balance for B (mol/min):} \qquad Q(C_{Bin} - C_B) - kC_AC_BV = 0 \qquad (3.13)$$
$$\text{Mass balance for C (mol/min):} \qquad Q(C_{Cin} - C_C) + kC_AC_BV = 0 \qquad (3.14)$$

Let us now calculate the energy balance for the fluid inside the reactor and for the fluid in the jacket. As the system is in a steady state, the accumulation term does not exist for both fluids. There is no chemical reaction inside the cooling jacket, so the conservation law applied to both fluids yields:

Energy balance for the fluid inside the reactor: $E - L + G/C = 0$

Energy balance for the cooling fluid: $E - L = 0$

As mentioned earlier, the generated or absorbed heat term (G/C) due to the chemical reaction is given by the reaction rate (mol/min) times $(-\Delta H)_R$ (J/mol):

$$G/C = kC_A C_B V(-\Delta H)_R (J/min)$$

In our case the reaction is exothermic $(\Delta H_R < 0)$. The minus sign in front of ΔH_R in the G/C term will guarantee that this amount of energy will be added to the system.

This reactor loses heat to the cooling jacket. Remembering that we adopt the convention that the convective heat term is added to the energy balance and that the gradient of temperature is represented by the surrounding temperature minus the temperature of the system being modeled, we obtain the following convective heat terms for the reactor and the jacket:

Reactor: Heat lost by convection to the cooling jacket: $U A (Tj - T)$ (J/min).
Jacket: Heat that the cooling jacket receives by convection: $U A (T - Tj)$ (J/min).

Finally, we need to consider the amount of energy that enters and leaves the system due to the flow of the fluids.

Fluid	Enters (J/min)	Leaves (J/min)
Inside the reactor	$Q \rho c_p T_{in}$	$Q \rho c_p T$
Cooling fluid	$Q_j \rho_j c_{pj} Tj_{in}$	$Q_j \rho_j c_{pj} Tj$

The energy balances for the reaction mixture and for the cooling fluid are shown in Eqs. (3.15) and (3.16). The mass balances Eqs. (3.12), (3.13), and (3.14) are also rewritten below:

Mass balance for A (mol/min): $Q(C_{Ain} - C_A) - kC_A C_B V = 0$ (3.12)

Mass balance for B (mol/min): $Q(C_{Bin} - C_B) - kC_A C_B V = 0$ (3.13)

Mass balance for C(mol/min): $Q(C_{Cin} - C_C) + kC_A C_B V = 0$ (3.14)

Energy balance for the reactor (J/min): $Q\rho c_p(T_{in} - T) + UA(Tj - T)$
$$+ kC_A C_B V(-\Delta H)_R = 0 \qquad (3.15)$$

Energy for the cooling fluid (J/min): $Q_j \rho_j c_{pj}(Tj_{in} - Tj) + UA(T - Tj) = 0$
$$(3.16)$$

The model for this reactor is represented by a system of five nonlinear algebraic equations that must be solved simultaneously. Considering the numerical values shown in Tables 3.1 and 3.2 for all parameters in the model and feed conditions, the concentrations and temperatures in a steady state can be obtained, as shown in Table 3.3. Observe in Table 3.3 that there are still reactants A and B in the reactor exit and that they are in the same amounts, due to the reaction stoichiometry.

Table 3.1 Parameters of the reaction mixture needed to simulate the continuous stirred tank reactor (CSTR)

ρ	c_p	A	V	k_0	F_A	R	ΔH_R	U
kg/m^3	J/kg K	m^2	m^3	m^3/mol min	J/mol	J/mol K	J/mol	J/min m^2 K
880	1750	5	40	8.2×10^5	48,500	8.314	$-72,800$	680

Table 3.2 Parameters of cooling jacket and feed conditions

c_{pj}	ρ_j	Q	Q_j	C_{Ain}	C_{Bin}	C_{Cin}	T_{in}	Tj_{in}
J/kg K	kg/m^3	m^3/min	m^3/min	mol/m^3	mol/m^3	mol/m^3	K	K
4180	1000	3	0.01	200	200	0	300	280

Table 3.3 Concentrations and temperatures in a steady state

C_A (mol/m^3)	C_B (mol/m^3)	C_C (mol/m^3)	T (K)	Tj (K)
49.5	49.5	150.5	307	282

If, for some reason, any parameter (such as inlet concentrations and temperatures, flow rates, etc.) undergoes variation, the concentrations and temperature of steady state presented in Table 3.3 will start changing with time until a new steady state is reached. The last example in this chapter (below) will address that.

Example 3.8 Mass and Energy Balance in a CSTR with a Cooling Jacket Operating in a Transient State

Let us imagine that, for some reason, the flow rates that enter and leave the reactor (Q), shown in Example 3.7, change simultaneously from 3 m^3/min to 4 m^3/min. The system will leave the steady state shown in Table 3.3; however, as the inlet and outlet valves operate at the same flow rate, the volume of the tank will remain the same and constant over time.

Assume that all parameters in the model, shown in Table 3.1, as well as the parameters for the cooling jacket, shown in Table 3.2, do not change. Assume also that the volume occupied by the cooling fluid inside the jacket is V_j (assume that $V_j = 0.032$ m^3). What would the profiles of the concentrations and temperatures be over time until a new steady state is reached?

Solution:

In order to model this system in a transient regime, the mass and energy balances need to be recalculated, but this time considering the accumulation terms that can be obtained as per Example 3.6 (see Table 3.4).

The accumulation terms in Table 3.4 need to be added to the mass and energy balances (Eqs. 3.12, 3.13, 3.14, 3.15, and 3.16). Observe that Table 3.4 represents the amount accumulated in Δt (min), so Eqs. (3.12), (3.13), (3.14), (3.15), and (3.16) need to be multiplied by Δt (min) to make the units compatible. Doing that and simplifying terms, the equations system that represents this reactor in a transient regime is obtained (Eqs. 3.17, 3.18, 3.19, 3.20, and 3.21)

Table 3.4 Accumulation terms to be added to the mass and energy balances to represent a transient regime

Balance	Accumulation term in Δt (min)	Unit
Reactant A	$V\dfrac{dC_A}{dt}\Delta t$	mol
Reactant B	$V\dfrac{dC_B}{dt}\Delta t$	mol
Product C	$V\dfrac{dC_C}{dt}\Delta t$	mol
Energy of the fluid inside the reactor	$V\rho c_p\dfrac{dT}{dt}\Delta t$	J
Energy of the cooling fluid	$V_j\rho_j c_{pj}\dfrac{dTj}{dt}\Delta t$	J

Mass balance for reactant A (mol): $V\dfrac{dC_A}{dt} = Q\,(C_{Ain} - C_A) - k\,C_A\,C_B V$ (3.17)

Mass balance for reactant B (mol): $V\dfrac{dC_B}{dt} = Q\,(C_{Bin} - C_B) - k\,C_A\,C_B V$ (3.18)

Mass balance for product C (mol): $V\dfrac{dC_C}{dt} = Q\,(C_{Cin} - C_C) + k\,C_A\,C_B V$ (3.19)

Energy Balance for the reactor (J): $V\rho c_p\dfrac{dT}{dt} = Q\rho c_p(T_{in} - T) + UA(Tj - T)$
$$+ kC_A C_B V(-\Delta H)_R$$
(3.20)

Energy balance for the cooling fluid (J): $V_j\rho_j c_{pj}\dfrac{dTj}{dt} = Q_j\rho_j c_{pj}(Tj_{in} - Tj)$
$$+ UA(T - Tj)$$
(3.21)

In order to solve this equation system, the initial conditions are needed. As the system was in a steady state before the perturbation, the initial conditions are the ones presented in Table 3.3. Solving this equation system and considering that $V_j = 0.032$ m^3, the profiles of the concentration and temperature over time until a new steady state is reached can be obtained (see Figs. 3.14 and 3.15).

Observe that Figs. 3.14 and 3.15 start from the steady state shown in Table 3.3, suffer variations over time due to the increase in the inflow and outflow rates, and finally reach another steady state.

With the concepts presented in this chapter, many lumped-parameter problems in chemical engineering can be modeled. Tools to numerically solve the models presented in this chapter will be presented in Chaps. 5 and 6, but, before studying that, Chap. 4 will show how to develop models for distributed systems, using the same recipe presented in Chap. 2.

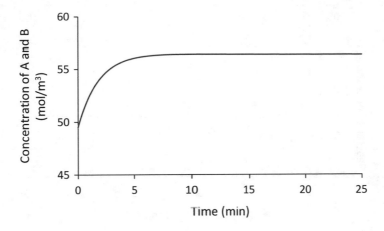

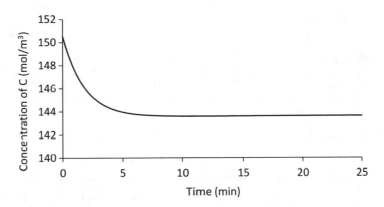

Fig. 3.14 Concentrations of A, B, and C over time for the continuous stirred tank reactor (CSTR) operating in a transient regime

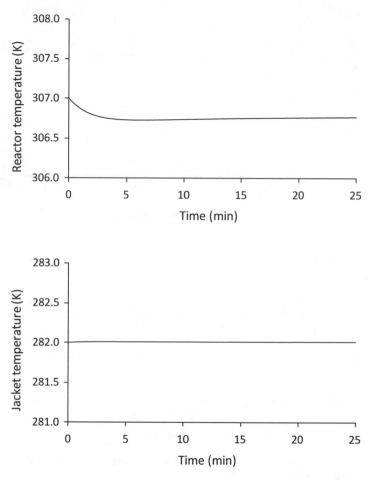

Fig. 3.15 Temperatures of the reactor and the jacket over time for the continuous stirred tank reactor (CSTR) operating in a transient regime

Proposed Problems

3.1) Imagine a perfectly stirred tank, shown in the Figure below, which contains 5 m^3 of a solution of HCl at a concentration of 0.01 kg/m^3. Two inlet valves are opened, both at a flow rate of 1 m^3/h, but one feeds a solution of HCl at 0.02 kg/m^3 and the other at 0.03 kg/m^3. At the same time, one outlet valve is opened and the solution of HCl leaves the tank at a flow rate equal to 2 m^3/h. Develop a model to obtain an ODE that represents the variation in the concentration of HCl in the tank over time. Define the initial condition needed to solve this equation.

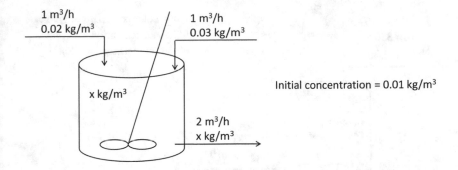

3.2) Without solving the ODE obtained in Proposed Problem 3.1, find the concentration of HCl inside the tank when a steady state is reached.

3.3) A perfectly stirred tank, shown as follows, initially with V m³ of water at T_0 K, is fed with $2Q$ m³/s of water at T_{in} K ($T_0 \neq T_{in}$). The tank has two outlet valves, each of them operating at a flow rate of Q m³/s. Assume that the water is incompressible and that there is no heat exchange with the environment (an adiabatic system).

(a) Develop a model to obtain an ODE to represent the variation in the water temperature over time.
(b) Without solving the ODE obtained in item (*a*) (above), find the temperature of the tank in a steady state.
(c) Assume that one of the outlet valves is closed. Write a model that represents this system.

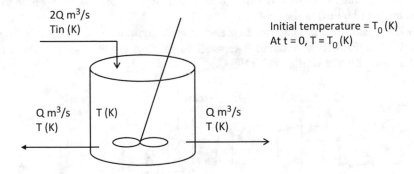

3.4) Imagine a perfectly stirred tank, shown as follows, in which there are simultaneous variations in concentration, temperature, and volume with time. Determine the ODE system that represents this tank, considering two situations: an adiabatic system; and heat exchange with the environment, which is at T_{env} (K). The global heat transfer coefficient is U (J/s m² °C) and the heat exchange area is A (m²). Create hypotheses to develop the model if needed.

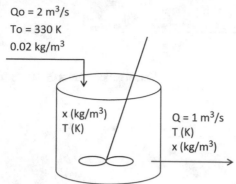

Qo = 2 m³/s
To = 330 K
0.02 kg/m³

Initial volume = 10 m³ of water
Initial concentration = 0 kg/m³
Initial temperature = 280K

x (kg/m³)
T (K)

Q = 1 m³/s
T (K)
x (kg/m³)

3.5) Imagine a hermetic cube of 0.001 m^3 (0.1 m $\times$ 0.1 m $\times$ 0.1 m) with ammonia at $10\ ^\circ$C inside it. Suddenly this cube is placed in an environment at a constant temperature of $30\ ^\circ$C, so the temperature of the ammonia inside the tube starts increasing. Consider that the global heat transfer coefficient is U (J/h m^2 $^\circ$C) and that the density and specific heat of the ammonia remain constant over time. Assume also that the temperature of the ammonia inside the cube is homogeneous (it does not depend on the position). Find the ODE that represents the variation in the temperature of the ammonia inside the cube over time. Define the initial condition used to solve this ODE.

3.6) Repeat problem 3.5 (above), but, instead of a cube, consider a sphere with a radius equal to R (m). Analyzing the model equations obtained from problems 3.5 and 3.6, find out the radius of the sphere to obtain the same temperature profile over time as that obtained in the cube.

3.7) Three tanks in series are used to preheat a multicomponent oil solution before it is fed into a distillation column for separation. This system is an adaptation of the system presented in Cutlip et al. (1998).

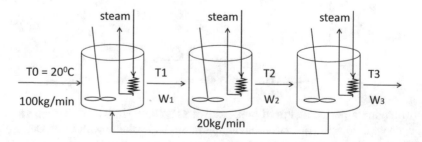

Each tank is initially filled with 1000 kg of oil at $20\ ^\circ$C. Saturated steam at a temperature of $250\ ^\circ$C condenses within a coil immersed in each tank. The oil is fed into the first tank at the rate $W = 100$ kg min^{-1} and overflows into the second and the third tanks at the same flow rate ($W = W_1 = W_2 = W_3$). The tanks are well mixed so the temperature inside the tanks is uniform. The specific heat (c_p) of the oil is 2.0 kJ/kg °C. For each tank, the rate at which heat is transferred to the oil from the stream coil is given by:

$$Q = UA(T_{stream} - T)$$

where $UA = 10$ kJ min^{-1} °C^{-1} is the product between the heat transfer coefficient and the area of the coil for each tank; T is the temperature of the oil in the tank (°C); and Q is the rate of heat transfer in kJ min^{-1}. The mass in each tank is constant, because the volume and the oil density do not vary. Assume there is no heat exchange with the environment.

(a) Find the ODEs system that represents the variation in the temperature over time for each tank. Define all initial conditions to solve the ODE system.
(b) Find the temperature in the three tanks in a steady state.

3.8) This problem is studied in Fogler (1999) and considers an isothermal CSTR from its startup to a steady state. Reactants A and B produce C and D according to the irreversible reaction: $A + B \xrightarrow{k} C + D$, in which the rate constant k is equal to 0.855 l/mol s. The reactor was initially fed with a solution containing product D at a concentration of 0.8 mol L^{-1} ($C_{D0} = 0.8$ mol L^{-1}). A solution with reactants A and B was added to the reactor at a flow rate of 5 L min^{-1} and at concentrations of A and B equal to 0.7 and 0.4 mol L^{-1}, respectively ($C_{A_1} = 0.7$ and $C_{B_1} = 0.4$ mol L^{-1}). The outlet volumetric flow rate is also 5 L min^{-1}, and the volume of liquid inside the reactor remains equal to 40 L over the entire reaction.

Find the ODE system that represents the concentrations of A, B, C, and D in the CSTR from startup to a steady state. Define all initial conditions to solve the equations. Create hypotheses for your model if needed.

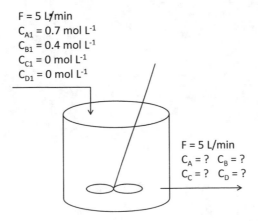

References

Bird, R.B., Steward, W.E., Lightfoot, E.N.: Transport Phenomena, 2nd edn. John Wiley & Sons, Inc., New York (2007)

Cutlip, M.B., Hwalek, J.J., Nuttall, H.E., Shacham, M., Brule, J., Widmann, J., Han, T., Finlayson, B., Rosen, E.M., Taylor, R.: A collection of 10 numerical problems in chemical engineering solved by various mathematical software packages. Compt. Appl. Eng. Educ. **6**, 169–180 (1998)

Davis, M.E., Davis, R.J.: Fundamentals of Chemical Reaction Engineering. McGraw-Hill Higher Education, New York (2003)

Fogler, H.S.: Elements of Chemical Reaction Engineering, 3rd edn. Prentice Hall, Upper Saddle River (1999)

Froment, G.F., Bischoff, K.B.: Chemical Reactor Analysis and Design, 2nd edn. Wiley, New York (1990)

Hill, C.G., Root, T.W.: Introduction to Chemical Engineering Kinetics and Reactor Design, 2nd edn. Wiley, Hoboken (2014)

Incropera, F.P., DeWitt, D.P., Bergman, T.L., Lavine, A.S.: Introduction to Heat Transfer, 5th edn. Wiley, New York (2006)

Kern, Q.D.: Process Heat Transfer. McGraw-Hill Book Company, New York (1950)

Levenspiel, O.: Chemical Reaction Engineering, 3rd edn. Wiley, New York (1999)

Missen, R.W., Mims, C.A., Saville, B.A.: Introduction to Chemical Reaction Engineering and Kinetics. Wiley, Hoboken (1999)

Welty, J.R., Wicks, C.E., Wilson, R.E., Rorrer, G.L.: Fundamentals of Momentum, Heat and Mass Transfer, 5th edn. Wiley, Hoboken (2007)

Chapter 4
Distributed-Parameter Models

In contrast to the previous chapter, which studied lumped-parameter problems, this chapter deals with examples in which variables such as concentration and temperature vary with position—a characteristic of distributed-parameter problems. As shown in Fig. 1.1, distributed-parameter problems can generate ordinary differential equations (ODEs) or partial differential equations (PDEs). In this chapter, we will see how mathematical models for distributed-parameter problems are developed, but the numerical solution of ODEs and PDEs will be presented only in Chaps. 6 and 7, respectively.

In this chapter, Sect. 4.1 gives some simple introductory examples needed to understand how to model distributed-parameter problems. Section 4.2 presents some concepts about transport by diffusion and models more complex systems. Finally, Sect. 4.3 presents some examples with variation in more than one spatial dimension.

4.1 Some Introductory Examples

Example 4.1 Assume that a fluid at 20 °C is fed into a cylindrical tube of length (L) 60 m and radius (R) 0.2 m at a rate (Q) of 4 m^3/h. Assume also that this tube exchanges heat with a jacket, whose temperature is 300 °C. Determine the axial profile of the temperature inside the tube. Consider that the system is in a steady state and there is no radial or angular variation of temperature inside the tube. Consider also that the thermal diffusion is not important in any direction (axial, radial, or angular).

Solution: According to the recipe presented in Chap 2, the first thing to do in order to model this tube is to define the control volume. In contrast to the examples

---

The original version of this chapter was revised. An erratum to this chapter can be found at https://doi.org/10.1007/978-3-319-66047-9_8

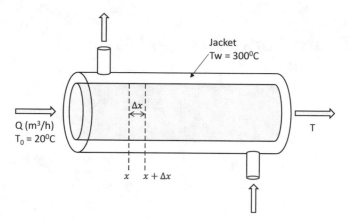

Fig. 4.1 Fluid flowing in a jacketed tube in a steady-state regime

presented in Chap. 3, the temperature of the fluid is not the same along the tube, so the entire tube cannot be considered as the control volume. In this way, we will consider a very small slice of the tube, of length Δx, and we will assume that in this small slice the fluid temperature is the same (see Fig. 4.1).

The conservation law will be applied to this small slice (control volume) in order to obtain the energy balance. This energy balance will be valid for any slice taken at any position inside the tube, so we can say that the energy balance obtained for the slice of tube shown in Fig. 4.1 will be valid for the entire tube.

In our problem, there is neither generation nor consumption of energy and the system is in a steady state, so the conservation law equation becomes:

$$E - L = 0 \tag{4.1}$$

Let us now define the energy that enters and leaves the control volume. Figure 4.1 shows that the fluid enters the control volume at position x and the amount of energy entering the control volume at x is given by:

$$Q\left(\frac{m^3}{h}\right)\rho\left(\frac{kg}{m^3}\right)c_p\left(\frac{J}{kg\ °C}\right)T(°C)$$

so:

Heat that enters in x:	$Q\,\rho\,c_p\,T$	(J/h)

$\tag{4.2}$

Be Careful Observe that we do not know the temperature of the fluid entering at position x, therefore we write a generic temperature T. Observe that 20 °C is the temperature of the fluid at position 0, and not at position x.

Figure 4.1 shows that the fluid leaves the control volume at $x + \Delta x$. In order to obtain the amount of energy that leaves the control volume, we use the concept of *infinitesimal variation of the dependent variable with the independent variable*, as shown below:

x (enters)	$x + \Delta x$ (leaves)	Dimension
$Q\rho c_p T$	$Q\rho c_p T + \dfrac{d(Q\rho c_p T)}{dx}\Delta x$	J/h

so:

$$\boxed{\text{Heat that leaves in } x + \Delta x: \qquad Q\rho c_p T + \frac{d(Q\rho c_p T)}{dx}\Delta x \qquad \text{(J/h)}} \qquad (4.3)$$

There is also heat exchange by convection with the jacket and, as defined in Chap. 3, this flow of energy can be expressed by:

Convection heat transfer rate $= U\,A\,(T_w - T)$, in which:

U = global coefficient of heat transfer (J/h m^2 °C)
$A = 2\,\pi\,R\,\Delta x$ = heat exchange area in the control volume (m^2)
T = temperature inside the control volume (°C)
T_w = temperature of the jacket (°C)

so:

$$\boxed{\text{Flow of Energy by Convection:} \qquad U2\pi R\Delta x\,(T_w - T) \qquad \text{(J/h)}} \qquad (4.4)$$

Observe that the heat exchange area by convection is the superficial area of the ring exchanging heat with the jacket. It is given by the perimeter of the ring $(2\pi R)$ times its length (Δx).

Using the expressions (4.2), (4.3), and (4.4) in the equation of conservation law (4.1) yields:

$$Q\rho c_p T - \left(Q\rho c_p T + \frac{d(Q\rho c_p T)}{dx}\Delta x \right) + U2\pi R\Delta x\,(Tw - T) = 0$$

or:

$$-\frac{d(Q\rho c_p T)}{dx} + U2\pi R\,(Tw - T) = 0 \qquad (4.5)$$

If we assume that ρ and c_p do not vary with temperature (and consequently with length), Eq. (4.5) becomes (remember $T_w = 300\ °C$):

$$\frac{dT}{dx} = \frac{2\pi R U}{Q\rho c_p}(300\text{-}T)$$

with the boundary condition that at $x = 0$, $T = 20\ °C$.

In order to analyze the temperature profile obtained for this system, let us assume the following numerical values: density $(\rho) = 900\ \text{kg/m}^3$, the specific heat of the fluid $(c_p) = 3000\ \text{J/kg °C}$, and the global coefficient of heat transfer $(U) = 60{,}000\ \text{J/h m}^2\text{°C}$. The profile of the temperature along the length can be seen in Fig. 4.2.

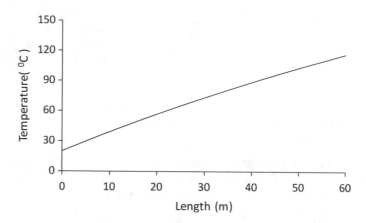

Fig. 4.2 Profile of temperature inside the jacketed tube

   Analyzing the temperature profile, it can be observed that the temperature of the fluid is 20 °C when it is fed into the tube, and it starts increasing along the length due to the heat exchange with the jacket, which is at 300 °C. Observe that at the end of the tube (60 m), the temperature of the fluid reaches 116 °C. If the tube were long enough, the temperature of the fluid would tend toward 300 °C, but it is impossible to reach temperatures higher than 300 °C inside the tube.

   Despite the fluid temperature changes along the length, the system is in a steady state because this axial profile of the temperature does not change over time.

Example 4.2 Let us imagine now that, for some reason, the temperature of the jacket in Example 4.1 changes instantaneously and abruptly from 300 °C to 200 °C. The system, which was in a steady state, suffers a modification, and the profile of the temperature inside the tube will change with time (the temperature will decrease) until the system reaches a new steady state. What would be the profile of the temperature of the fluid inside the tube along the length and over time until the new steady state is reached? Solve this problem, assuming neither radial nor angular temperature profiles, and no heat diffusion in any direction inside the tube.

Solution: If the temperature of the jacket changes abruptly from 300 °C to 200 °C, the profile of the temperature shown in Fig. 4.2 will be modified. The temperature profile will still start at 20 °C because the fluid is fed into the tube at this temperature, but the fluid temperature will increase less because the jacket is 100 °C colder. One can imagine that the axial profile of the temperature inside the tube will start as shown in Fig. 4.2 (at time = 0), but it will change over time until a new steady state is reached.

   The modeling of this system will generate a PDE because the temperature will change along the tube and over time.

   The same slice of the tube shown in Fig. 4.1 will be considered as the control volume. As there is temperature variation with time, the accumulation term must be considered, and the conservation law equation becomes:

$$E - L = A$$

The expressions to represent the amount of energy that enters at x and leaves at $x + \Delta x$, as well as the convection term, are calculated as per Example 4.1 and are rewritten below:

Heat that enters in x (J/s):	$Q \rho c_p T$	(4.6)
Heat that leaves in $x + \Delta x$ (J/s):	$Q \rho c_p T + \dfrac{d(Q \rho c_p T)}{dx} \Delta x$	(4.7)
Flow of Energy by Convection (J/s):	$U 2 \pi R \Delta x (T_w - T)$	(4.8)

Besides these three terms, we need to calculate the amount of energy accumulated in the control volume. We will do that as was done in the previous chapter, i.e., by calculating the amount of energy accumulated in a very short period of time Δt, which is the energy at time $t + \Delta t$ minus the energy at time t (see below).

t	$t + \Delta t$	Dimension
$V \rho c_p T$	$V \rho c_p T + \dfrac{d(V \rho c_p T)}{dt} \Delta t$	J

Once again, we assume that ρ and c_p do not change with temperature and consequently with time, so the accumulation term becomes:

$$\text{Accumulation term} = V \rho c_p \frac{dT}{dt} \Delta t \qquad \text{(J)}$$

Observe that the control volume is a small cylinder with an area of the base equal to πR^2 and height equal to Δx, so the volume, in the accumulation term, is the control volume $(V = \pi R^2 \Delta x)$.

Since the accumulation term represents the amount of energy that is accumulated in a period of time Δt, we must also consider the energy that enters and leaves the control volume in this same period of time, so we must multiply expressions (4.6), (4.7), and (4.8) by Δt. The energy balance in the period Δt becomes:

$$Q \rho c_p T \Delta t - \left(Q \rho c_p T + \frac{d(Q \rho c_p T)}{dx} \Delta x \right) \Delta t + U 2 \pi R \Delta x (Tw - T) \Delta t = V \rho c_p \frac{dT}{dt} \Delta t$$

or:

$$-Q \rho c_p \frac{dT}{dx} \Delta x \Delta t + U 2 \pi R \Delta x (Tw - T) \Delta t = \pi R^2 \Delta x \, \rho c_p \frac{dT}{dt} \Delta t$$

Observe that Δt and Δx appear in all terms of the energy balance and can be simplified. In fact, this simplification must always happen when developing a model. If at this point we were not able to cancel all Δ's from our balance, it is because we made some sort of mistake, and our model has to be double checked.

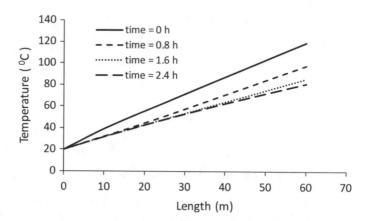

Fig. 4.3 Profiles of fluid temperature along the length over time

Also observe that there are two independent variables (x and t), so the symbol d must be changed to ∂. So the energy balance becomes:

$$\frac{\partial T}{\partial t} = \frac{2U}{R\rho c_p}(Tw - T) - \frac{Q}{\pi R^2}\frac{\partial T}{\partial x} \tag{4.9}$$

This PDE needs one boundary condition (related to the length) and one initial condition (related to time). The initial condition can be obtained remembering that, at the beginning (at $t = 0$ h), the profile of the temperature inside the tube is the one shown in Fig. 4.2, because the system operated like that before the change in the jacket temperature. The boundary condition is obtained remembering that the temperature of the fluid fed into the tube is 20 °C, so at $x = 0$, $T = 20$ °C.

After solving Eq. (4.9) using these two initial/boundary conditions, the temperature profiles along the length and over time, as shown in Fig. 4.3, can be obtained.

One can observe that the axial profile of the temperature varies over time, with greater variations in the beginning (see the difference in the profiles from 0 to 0.8 h) and minor variations as time goes on (see the small difference in the curves from time 1.6 h to 2.4 h) until a new steady state is reached, when the axial profile of the temperature does not change anymore with time. In our case, after 2.4 h the temperature profile along the length remains the same, so we can say that the new steady state was reached after 2.4 h.

Example 4.3 This problem is adapted from the book *Process Heat Transfer*, by Q. D. Kern (1950), and develops a model to design bitubular heat transfer. Let us consider now two concentric tubes, as shown in Fig. 4.4, with benzene flowing through the internal tube and toluene flowing through the annulus. The two fluids flow in parallel, and the system is in a steady state. The benzene and toluene are fed at rates of 9820 lb/h (W_{ben}) and 6330 lb/h (W_{tol}), respectively. These concentric tubes are used to increase the temperature of benzene from 60 °F to 100 °F and reduce the temperature of toluene from 170 °F to 110 °F. Assume that the toluene

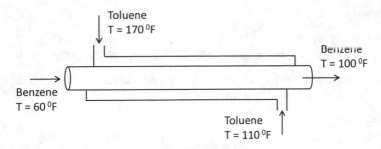

Fig. 4.4 Double-pipe heat exchanger operating with parallel flow

does not exchange heat with the environment, just with the benzene, so that the heat exchange occurs through the wall of the internal tube, which has a diameter of 1.25 in (ignore the thickness of the tube). What is the required length of the tubes to make the necessary thermal exchange? Let us assume that the c_p values of benzene (c_{pben}) and toluene (c_{ptol}) are 0.425 Btu/(lb °F) and 0.440 Btu/(lb °F), respectively, and that these values do not change significantly with temperature.

Solution: The solution to this problem is similar to the solution to problem 4.1, but in this case, it is necessary to make the energy balance for both fluids. We will first present a solution similar to what we have done so far, but in the sequence, some assumptions will be made, in order to bring the solution of this problem closer to what is done in the design of heat exchange equipment.

   Heat exchangers are very common in chemical industries, and they are very useful to exchange heat between different streams in a plant. This equipment can assume many different geometries, and the one considered in this example is the simplest (a bitubular heat exchanger with parallel flow).

   To model this system, we must define a control volume. As this is a distributed-parameter problem (the temperature changes along the length), an increment Δx is considered and the energy balance for both fluids is calculated considering this small volume. The amount of heat that enters and leaves the control volume is presented as follows:

	x	$x + \Delta x$	Dimension
Benzene	$W_{ben}cp_{ben}T_{ben}$	$W_{ben}cp_{ben}T_{ben} + \dfrac{d(W_{ben}cp_{ben}T_{ben})}{dx}\Delta x$	Btu/h
Toluene	$W_{tol}cp_{tol}T_{tol}$	$W_{tol}cp_{tol}T_{tol} + \dfrac{d(W_{tol}cp_{tol}T_{tol})}{dx}\Delta x$	Btu/h

   The amount of heat exchanged by convection is written in the same way as was done before, keeping in mind the *sign convention* presented in Example 3.4.

Heat exchanged by convection (Btu/h)	
Benzene	$U\,A\,(T_{tol} - T_{ben})$
Toluene	$U\,A\,(T_{ben} - T_{tol})$

in which:

U = global coefficient of heat transfer (assume that $U = 0.8$ Btu/(h in^2 °F))
$A = 2\pi R \Delta x$ = superficial area for the heat exchange in the control volume (in^2)

The conservation law for benzene and toluene yields $E - L = 0$, and the energy balance for both fluids can be written as:

Benzene:

$$W_{ben}cp_{ben}T_{ben} - \left(W_{ben}cp_{ben}T_{ben} + \frac{d(W_{ben}cp_{ben}T_{ben})}{dx}\Delta x\right) + U2\pi R\Delta x(T_{tol} - T_{ben}) = 0$$

Toluene:

$$W_{tol}cp_{tol}T_{tol} - \left(W_{tol}cp_{tol}T_{tol} + \frac{d(W_{tol}cp_{tol}T_{tol})}{dx}\Delta x\right) + U2\pi R\Delta x(T_{bez} - T_{tol}) = 0$$

After simplifying terms, and remembering that the internal tube has diameter $(D=2R)$ equal to 1.25 in, we obtain:

Benzene: $W_{ben}cp_{ben}\dfrac{dT_{ben}}{dx} = U1.25\pi(T_{tol} - T_{ben})$,   at $x = 0, T_{ben} = 60°$F

Toluene: $W_{tol}cp_{tol}\dfrac{dT_{tol}}{dx} = U1.25\pi(T_{ben} - T_{tol})$,   at $x = 0, T_{tol} = 170°$F

Considering numerical values for all parameters ($W_{ben} = 9820$ lb/h, $W_{tol} = 6330$ lb/h, $cp_{ben} = 0.425$ Btu/(lb °F), $cp_{tol} = 0.440$ Btu/(lb °F), and $U = 0.8$ Btu/(h in^2 °F)), this system of two ODEs can be solved numerically to generate temperature profiles for benzene and toluene along the length of the tubes, as shown in Fig. 4.5.

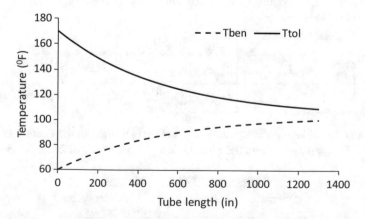

Fig. 4.5 Profiles of temperature inside the bitubular heat exchanger

One can observe that the benzene and toluene reach the desired temperatures ($T_{ben} = 100$ °F and $T_{tol} = 110$ °F) when the length of the tubes is around 1290 inches (or 110 feet).

A second approach presented in the sequence is more frequently used in the design of heat exchangers because many simplifying hypotheses are proposed to reach the numerical solution easier and faster. Typically, the design of this equipment means finding the size (area) of a heat exchanger able to make the desired changes in the temperature in hot and cold fluids. Assuming there is no heat loss to the environment, the amount of heat received by the cold fluid is equal to that lost by the hot fluid, and this amount is equal to the energy exchanged by convection between the two fluids. The amount of energy exchanged by convection varies along the length of the tube because the difference in temperature between the hot and cold fluids (the driving force) varies along the length (see Fig. 4.5).

In the design of heat transfer equipment, the logarithmic mean temperature difference (LMTD) is considered to simplify numerical solution. The LMTD is an average of the difference in temperature between hot and cold fluids along the entire equipment length (the average driving force along the entire equipment length) and is given by:

$$\text{LMTD} = \frac{\Delta T_A - \Delta T_B}{\ln\left(\frac{\Delta T_A}{\Delta T_B}\right)} = \frac{\Delta T_A - \Delta T_B}{\ln \Delta T_A - \ln \Delta T_B}$$

in which ΔT_A is the difference in temperature between the two streams at the heat transfer end where the hot fluid is fed in (in this case, $170 - 60 = 110$ °F) and ΔT_B is the difference in temperature between the two streams at the other end (in this case, $110 - 100 = 10$ °F). In our case, the LMTD is 41.7 °F.

The total amount of heat received by the cold fluid and lost by the hot fluid is $W_{ben} cp_{ben} (100 - 60) = W_{tol} cp_{tol} (170 - 110) = 1.67 \times 10^5$, and this amount is equal to the heat exchanged by convection between the two fluids ($U \times A \times LMTD$), so $U \times A \times LMTD = 1.67 \times 10^5$. In this way, $A = 1.67 \times 10^5 / (U \times LMTD)$. Considering numerical values for U (0.8 Btu / h in^2 °F) and $LMTD$ (41.7 °F), we obtain $A = 5006$ in^2.

The heat exchange area of the tube ($A = 5006$ in^2) can be written as πDL, in which D is the diameter of the tube (1.25 in) and L is its length. In this way, the length of the tube can be found as $L = A/\pi D$, or $L = 1275$ in.

Observe that the value obtained using this simplified calculation ($L = 1275$ in) is only 1.2% lower than the one obtained when the system of two ODEs are solved together ($L = 1290$ in). For this reason, a project for heat exchanger equipment is usually done as per the second approach; however, the recipe presented in this book could also be used to design heat exchangers.

After these three simple introductory examples on distributed-parameter problems, Sect. 4.2 will revisit some concepts on mass, energy, and momentum transfer by diffusion, which are needed to model more complex systems. The idea is just to show a few pieces of information needed to model systems in which transfers by

58   4 Distributed-Parameter Models

diffusion are relevant. The reader can find more details on this subject in specific literature dealing with transfer phenomena (Bird et al. 2007; Welty et al. 2007; Bergman et al. 2011, just to mention a few).

4.2 Concepts About Transfer by Diffusion

All flow rates (mass, heat, and momentum) are directly proportional to a driving force and inversely proportional to the resistance:

$$\text{Flow Rate } \alpha \; \frac{\text{Driving Force}}{\text{Resistance}}$$

We used this concept in Chap. 3 (Sect. 3.2) for heat flow, but it is also valid for mass and momentum. This section will revisit some concepts about diffusive transport and will point out some analogies among transfers of heat, mass, and momentum.

4.2.1 Diffusive Transport of Heat

Imagine a solid cube of 1 m^3 initially at 50 °C (Fig. 4.6a). Two opposite sides of the cube are fixed to surfaces that are at 70 °C and 30 °C (assume that both temperatures do not vary over time) and all other faces are insulated, so there is energy flow only in the x direction.

One can imagine that as the energy starts flowing from the hotter face to the colder face, the internal profile of the temperature inside the cube starts changing until the system reaches a steady state. Figure 4.6b shows an example of how the axial profiles of the temperature inside the cube could change over time. Since there is no heat exchange with the environment, in a steady state the temperature profile inside the cube is given by a straight line.

The energy flow inside the cube is due only to heat diffusion (heat conduction from molecule to molecule). Fourier observed that the flow of energy by diffusion can be expressed by:

$$\left(\frac{q}{A}\right)_x = -k\left(\frac{dT}{dx}\right) \tag{4.10}$$

in which (using international units):

q = energy flow (J/s)
A = cross-sectional area from where the energy flows (m^2)
$(q/A)_x$ = energy flux in the x direction (J/s m^2)

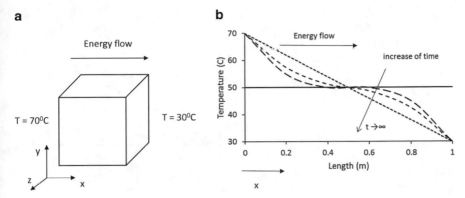

Fig. 4.6 Diffusive transport of heat. (**a**) Solid cube, initially at 50 °C, with four insulated faces and two opposite faces at 70 °C and 30 °C. (**b**) Profiles of temperature along the length of the cube over time

k = thermal conductivity (J/s m °C)
dT/dx = temperature gradient (°C/m)

Observe that the energy flux $(q/A)_x$ of the Fourier equation is directly proportional to the driving force (dT) and inversely proportional to the resistance (dx/k), as shown in Sect. 3.2.

The negative sign of Eq. (4.10) is due to the orientation of q, dx, and dT. Observe in Fig. 4.6b that dx is positive and dT is negative, making dT/dx negative. Since the energy flux $(q/A)_x$ is positive (it grows in the same orientation as x), the minus sign is necessary to make Eq. (4.10) coherent.

4.2.2   Diffusive Transport of Mass

Imagine a cubic box of 1 m³, open at the top with only air inside it. At some point, ethylene gas at a constant concentration $C = 1$ mol/m³ starts blowing above the box, as shown in Fig. 4.7a. Imagine that, by some mechanism that is chemically possible, the concentration of ethylene at the bottom of the box is always zero.

At the beginning, the concentration of ethylene inside the box is zero, but, as the ethylene starts blowing, there will be a diffusive mass flow of ethylene from the region with a higher concentration to the region with a lower concentration of ethylene, so in this case, the *driving force* is the *difference in concentration*. (In fact, the driving force for the mass transfer is the chemical potential, which includes pressure and thermal energies as well as the energy due to the molecular interaction. Since, in this system, the pressure and temperature are the same for all of the system, the chemical potential is the difference in concentration.)

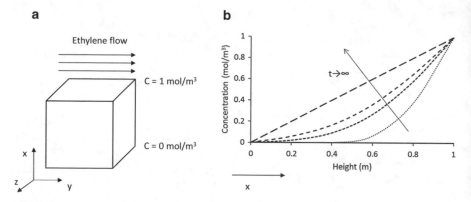

Fig. 4.7 Diffusive transport of mass. (**a**) Cubic box initially containing air, with ethylene concentrations at the bottom and at the top equal to 0 and 1 mol/m³, respectively. (**b**) Profiles of ethylene concentration along the cube height over time

The diffusive mass flow will generate axial profiles of ethylene concentration inside the box, which will vary over time until a steady state is reached (see Fig. 4.7b). In this example, we assume there is no profile of the concentration along y and z.

In 1855, Fick experimentally obtained Eq. (4.11) to represent the molar flow of some compound (say, A) by diffusion.

$$\left(\frac{J_A}{A}\right)_x = -D\left(\frac{dC_A}{dx}\right) \qquad (4.11)$$

in which (using international units):

J_A = molar flow of component A (mol/s)
A = cross-sectional area from where the molecules of component A flow (m²)
$(J_A/A)_x$ = molar flux of component A in the mass flow direction (mol/m² s)
D = diffusion coefficient or diffusivity of component A (m²/s)
dC_A/dx = concentration gradient for component A (mol/m³ m)

In this case it is also observed that the molar flux of component A $(J_A/A)_x$ in Fick's law is directly proportional to the driving force (dC_A) and inversely proportional to the resistance (dx/D).

Analogous to Eq. (4.10), the negative sign in Fick's law is necessary to make Eq. (4.11) coherent with the orientation of J_A, dx, and dC_A.

4.2.3   Diffusive Transport of Momentum

Fluids can be classified as Newtonian and non-Newtonian, but definitions and examples of the different types of fluids are not within the scope of this book and

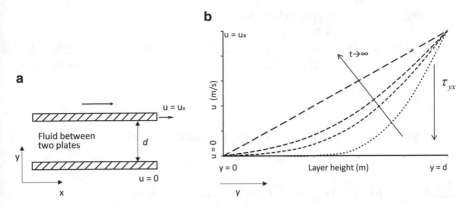

Fig. 4.8 Diffusive transport of momentum. (**a**) Thin layer of a Newtonian fluid between two long plates. (**b**) Profiles of velocity along the layer height over time

can be found in literature related to the mechanics of fluids and transport phenomena (Bird et al. 2007; Welty et al. 2007; Fox et al. 2012; etc).

From a didactic point of view, the example presented in this section considers a Newtonian fluid. In our example, imagine a thin layer of a fluid between two long plates, as shown in Fig. 4.8a. The thickness of the fluid layer is d(m). Initially ($t = 0$) the system is at rest, and after that, the upper plate starts moving in the positive direction of x at velocity u equal to u_x (m/s). Consider that the lower plate stays still ($u = 0$).

At the beginning ($t = 0$) the fluid velocity is zero. When the movement starts, the molecules of the fluid in contact with the plates assume the same velocity as the plates and, due to the frictional forces, a velocity profile is formed, as shown in Fig. 4.8b. As time goes on, more and more fluid is drawn toward the moving plate.

Newton observed that in laminar flow, when a steady state is reached, there is a linear velocity profile inside the fluid layer (Newton's law of viscosity), as can be observed in Fig. 4.8b.

If the force applied to the upper plate is "F" and the area of the upper plate is "A", the ratio "F/A" is known as *shear stress*, and it is equal in magnitude to the momentum flux. Physically, the momentum flux is the transfer of momentum through the fluid from a region with higher velocity to another region with lower velocity, and can be represented by τ_{yx}, in which y is the direction of the transfer of momentum and x is the direction of the movement velocity.

In Newton's law of viscosity, the momentum flux (τ_{yx}) is directly proportional to a driving force (du) and inversely proportional to the resistance dy/μ, as can be seen in Eq. (4.12):

$$\boxed{\frac{F}{A} = \tau_{yx} = -\mu\left(\frac{du}{dy}\right)} \tag{4.12}$$

in which (using international units):

F = force applied to the upper plate (N)

A = area of the upper plate (m^2)

τ_{yx} = shear stress (momentum flux) in which y represents the normal component of the plane of action of the shear stress and x is parallel to the velocity of the plate (N/m^2)

μ = viscosity of the fluid (N s/m^2)

du/dy = gradient of velocity (1/s)

For the same reason mentioned before, the minus sign in Eq. (4.12) is needed.

4.2.4 Analogies Among All Diffusive Transports

Observe that the same behavior is observed for the diffusive transfers of heat, mass, and momentum. Table 4.1 summarizes the analogies among the three kinds of diffusive transport presented before.

Table 4.1 Analogies among kinds of diffusive transport

	Heat	Mass	Momentum
Law	Fourier	Fick	Newton
Equation	$\left(\dfrac{q}{A}\right)_x = -k\left(\dfrac{dT}{dx}\right)$	$\left(\dfrac{J_A}{A}\right)_x = -D\left(\dfrac{dC_A}{dx}\right)$	$\dfrac{F}{A} = \tau_{yx} = -\mu\left(\dfrac{du}{dy}\right)$
Driving force	dT	dC_A	du
Resistance	dx/k	dx/D	dy/μ

4.2.5 Examples Considering the Diffusive Effects on the Modeling

The examples presented in this section take into account problems in which the diffusive effects are important and cannot be neglected. From a didactic point of view, in this section, variations in only one direction will be considered. Examples 4.4 and 4.7 are analogous to the schemes presented in Sects. 4.2.1 and 4.2.2, so it will be possible to visualize the mathematical models that generate Figs. 4.6b and 4.7b.

Example 4.4 Imagine a cylindrical metal bar with a length (L) of 1 m and a radius (R) equal to 0.03 m, initially at 50 °C (Fig. 4.9). At some point, the two ends of the bar are fixed to walls that are at 70 °C and 30 °C (assume that the temperature of the walls does not vary with time). Imagine that this bar is insulated, so there is no heat transfer by convection between the bar and the environment. This system is very similar to the one shown in Fig. 4.6a. In this example, we want to know how long it takes to reach a steady state and what the internal profiles of the temperature will be until then.

Fig. 4.9 Heat conduction in an insulated cylindrical bar

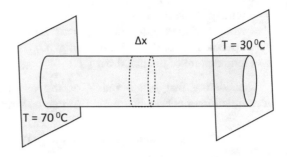

Solution: As this is a distributed-parameter problem (there is an axial profile of the temperature), a slice of the cylinder must be considered as the control volume, as depicted in Fig. 4.9, whose volume is $\pi R^2 \Delta x$.

The conservation law applied to this control volume yields: $E - L = A$.

Observe that there is neither generation nor consumption of energy in this case, and there is no heat transfer by convection with the environment. The accumulation term must be considered because the axial profile of the temperature will change with time until a steady state is reached.

The flow of energy (q) inside the cylinder occurs by conduction (from molecule to molecule of the metal) due to the difference in temperature (driving force) and can be represented by the Fourier law (Eq. 4.10), rewritten below:

$$q = -k\,A\,\frac{dT}{dx} \tag{4.13}$$

The amount of energy that enters (at x) and leaves (at $x + \Delta x$) the control volume by conduction is shown as follows:

	x (enters)	$x + \Delta x$ (leaves)
Energy flow by conduction (J/s)	$-k\pi R^2 \dfrac{dT}{dx}$	$-k\pi R^2 \dfrac{dT}{dx} + \dfrac{d}{dx}\left(-k\pi R^2 \dfrac{dT}{dx}\right)\Delta x$

in which (using international units):

k = thermal conductivity (J/s m °C)
πR^2 = cross-sectional area from where the energy flows (m^2)
dT/dx = temperature gradient (°C/m)

The accumulation of energy (in joules) inside the control volume in a period Δt (s) can be obtained by the difference in energy at the times $t + \Delta t$ and t.

	t	$t + \Delta t$
Energy (J)	$\rho V c_p T$	$\rho V c_p T + \dfrac{d(\rho V c_p T)}{dt}\Delta t$

in which (using international units):

ρ = density of the metal bar (kg/m^3)
V = volume of the control volume = $\pi R^2 \Delta x$ (m^3)
c_p = specific heat of the metal bar (J/kg °C)
T = temperature of the metal bar (°C)

Considering that ρ, c_p, and V of the metal do not change over time, the accumulation of energy in a period Δt can be written as:

$$\text{Accumulation of Energy } \Delta t(\text{J}) = \rho c_p \pi R^2 \Delta x \frac{dT}{dt} \Delta t$$

Since the amount of energy entering and leaving the control volume is given in J/s and the accumulation of energy is in J, the energy flow must be multiplied by Δt to make the units compatible. Doing that, the energy balance can be written as follows:

$$-k\pi R^2 \frac{dT}{dx} \Delta t - \left[-k\pi R^2 \frac{dT}{dx} + \frac{d}{dx}\left(-k\pi R^2 \frac{dT}{dx} \right) \Delta x \right] \Delta t = \rho c_p \pi R^2 \Delta x \frac{dT}{dt} \Delta t \quad (4.14)$$

Simplifying terms and considering that the thermal conductivity of the metal remains constant, the following PDE is obtained:

$$\frac{\partial T}{\partial t} = \frac{k}{\rho c_p} \frac{\partial^2 T}{\partial x^2} \quad (4.15)$$

To solve this PDE, three conditions are needed: two related to space and one related to time. Remember that at the beginning (at t = 0), the entire bar is at 50 °C, and after that, the two ends of the bar are kept at 70 °C and 30 °C (left and right). So the three conditions are:

$$\text{At} \quad t = 0\,\text{h}, \quad T = 50°\text{C}, \quad \text{for} \quad 0 \leq L \leq 1\,\text{m}$$

$$\text{At} \quad x = 0\,\text{m}, \quad T = 70°\text{C}, \quad \text{for} \quad t > 0\,\text{h}$$

$$\text{At} \quad x = 1\,\text{m}, \quad T = 30°\text{C}, \quad \text{for} \quad t > 0\,\text{h}$$

Considering numerical values (k = 398.2 J/s m °C, c_p = 386.3 J/kg °C, and ρ = 8933 kg/m^3), the PDE can be solved numerically and can generate axial profiles of the temperature over time, as shown in Fig. 4.6b. For a metal with the properties considered in this example, after 10 min, the internal profile of the temperature inside the bar almost does not change anymore over time.

If one would like to know only the axial profile of the temperature in a steady state, the energy balance presented in Eq. (4.15) could be simplified because the accumulation term would not be used, so the final energy balance would be:

$$\frac{d^2 T}{dx^2} = 0 \quad (4.16)$$

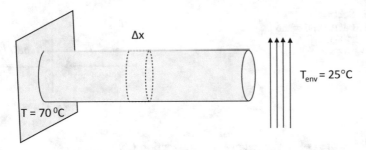

Fig. 4.10 Insulated cylinder bar exchanging heat with the environment at only one of its ends

with the boundary conditions:

At $x = 0$ m, $T = 70\,°C$
At $x = 1$ m, $T = 30\,°C$

Equation (4.16) can be easily integrated to generate $T = -40x + 70$, which is the linear profile of the temperature in a steady state, shown in Fig. 4.6b.

Example 4.5 Now imagine that the cylindrical metal bar of the in Example 4.4 is initially at 50 °C, but, this time, only one end of the bar is fixed to a wall at 70 °C (a constant temperature) and the other exchanges heat with the environment, which is at 25 °C (see Fig. 4.10). Consider that the rest of the cylindrical metal bar is insulated (no heat exchange with the environment). Do the modeling of this system again to obtain how the internal profile of the temperature changes over time.

Solution: Since the lateral of the control volume remains insulated, the conservation law applied to this control volume will generate the same energy balance represented by Eq. (4.15), rewritten below.

$$\frac{\partial T}{\partial t} = \frac{k}{\rho c_p} \frac{\partial^2 T}{\partial x^2} \tag{4.15}$$

The conditions at $t = 0$ and $x = 0$ also stay the same, but the boundary condition at $x = 1$ m is different this time. At this end, the heat flowing by conduction from inside the bar to the end (at $x = 1$) is equal to the heat that leaves the bar (at $x = 1$) by convection (exchanging heat with the environment), or:

$$\text{At } x = 1\,\text{m:} \qquad -kA\frac{dT}{dx} = hA'(T - T_{env}) \tag{4.17}$$

In which (using international units):

k = thermal conductivity (J/s m °C)
$A = \pi R^2$ = cross-sectional area from where the energy flows by conduction (m^2)
dT/dx = temperature gradient (°C/m)
h = global coefficient of heat transfer by convection (J/s m^2 °C)

Fig. 4.11 Scheme of
energy flow at the end of the
bar (at $x = 1$ m)

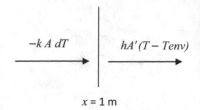

$A' = \pi R^2$ = area of heat exchange by convection (m^2)
T_{env} = environmental temperature (°C)
T = temperature of the bar (in this case, at $x = 1$ m) (°C)

Figure 4.11 shows a scheme of the flow of energy at $x = 1$ m.

It is important to note that the heat represented by both terms in Eq. (4.17) has the same magnitude and flows toward the same orientation, so both must have the same sign (positive or negative). The conduction term is positive because, in this example, dT is negative, dx is positive, and the Fourier expression gives a minus sign. In this way, the convective term must also be positive, therefore we used $(T - T_{env})$ to represent the driving force $(T > T_{env})$. (Observe that the *simple convention* presented in Example 3.4, for the convective heat transfer in order to build the model, is not used for the boundary condition).

Observe in Eq. 4.17 that the cross-sectional area from where the energy flows by conduction at $x = 1$ m (A) is equal to the area from where the bar exchanges heat with the environment by convection (A'), so both terms can be simplified.

So the three conditions needed to solve Eq. (4.15) are:

$$\text{At } t = 0 \text{ h}, \ T = 50°\text{C}, \ \text{for } 0 \leq L \leq 1 \text{ m}$$

$$\text{At } x = 0 \text{ m}, \ T = 70°\text{C}, \ \text{for } t > 0 \text{ h}$$

$$\text{At } x = 1 \text{ m}, \ \frac{dT}{dx} = -\frac{h}{k}(T - T_{env}), \ \text{for } t > 0 \text{ h}$$

Solving Eq. (4.15) with this new set of initial/boundary conditions and assuming that $h = 300$ J/min m^2 °C, the axial profiles of the temperature over time shown in Fig. 4.12 can be obtained.

One can observe that Fig. 4.12 shows temperature profiles very different from the ones presented in Fig. 4.6b, because one of the ends of the bar exchanges heat with the environment.

Example 4.6 Now, imagine that the same bar presented in Example 4.5 is not insulated anymore. How would the axial profiles of the temperature change inside the bar until a steady state is reached?

Solution: For a noninsulated bar, the conservation law applied to the control volume must consider the heat transfer by convection between the bar and the environment:

$$\text{Convective term: } h \, 2\pi R \Delta x \, (T_{env} - T) \tag{4.18}$$

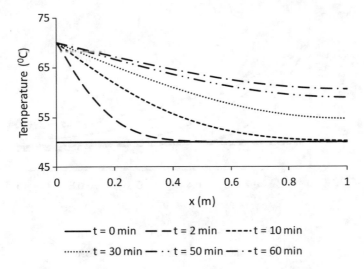

Fig. 4.12 Axial profiles of temperature over time when one end of the bar changes heat with the environment

in which (using international units):

h = global coefficient of heat transfer by convection (J/s m^2 °C)
$2\pi R \Delta x$ = superficial area from where there is heat exchange by convection between the control volume and the environment (m^2)
T_{env} = environmental temperature (°C)
T = temperature in the control volume (°C)

Adding the expression (4.18) to the energy balance (4.14) used in Examples 4.4 and 4.5 (for an insulated bar), we obtain Eq. (4.19). Observe that the energy balance represented by Eq. (4.14) considers the amount of heat that enters, leaves, and accumulates in a period Δt; therefore the expression (4.18) is also multiplied by Δt:

$$-k\pi R^2 \frac{dT}{dx}\Delta t - \left[-k\pi R^2 \frac{dT}{dx} + \frac{d}{dx}\left(-k\pi R^2 \frac{dT}{dx}\right)\Delta x\right]\Delta t$$
$$+ 2h\pi R\Delta x(T_{env} - T)\Delta t = \rho c_p \pi R^2 \Delta x \frac{dT}{dt}\Delta t \tag{4.19}$$

After simplifying terms, Eq. (4.19) becomes:

$$\rho\, c_p \frac{\partial T}{\partial t} = k \frac{\partial^2 T}{\partial x^2} + \frac{2h}{R}(T_{env} - T) \tag{4.20}$$

If Eq. (4.20) is written in terms of the diameter of the metal bar (D), and not in terms of the radius (R), we can obtain, after rearranging terms:

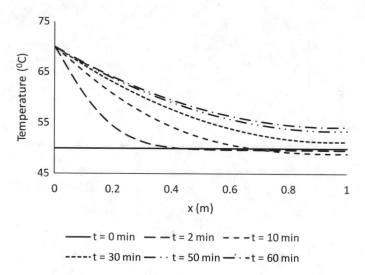

Fig. 4.13 Axial profiles of temperature over time when the bar changes heat with the environment

$$\frac{\partial T}{\partial t} = \left(\frac{k}{\rho c_p}\right)\frac{\partial^2 T}{\partial x^2} + \frac{4h}{D\rho c_p}(T_{\text{env}} - T) \tag{4.21}$$

The conditions needed to solve Eq. (4.21) are the same as those used in Example 4.5:

At $t = 0$ h, $T = 50\,^\circ$C, for $0 \leq L \leq 1$ m

At $x = 0$ m, $T = 70\,^\circ$C, for $t > 0$ h

At $x = 1$ m, $\dfrac{dT}{dx} = -\dfrac{h}{k}(T - T_{\text{env}})$, for $t > 0$ h

After solving Eq. (4.21), the axial profiles of the temperature in the bar over time are obtained, and are shown in Fig. 4.13.

In comparison with Fig. 4.12, we can observe that the temperature of the bar is lower, and the decay is faster, as expected. The velocity of the decay and the temperature will depend on the numerical values of the parameters in the model, such as h, c_p and ρ.

Example 4.7 Imagine a cylinder vase 1 cm in radius (R) and 5 cm high (L), open at the top with only air inside it. This cylinder is placed in an atmosphere containing a certain gas A in a concentration (C_A) equal to 1 mol/m^3 (see Fig. 4.14). As the concentration of A outside the cylinder vase is higher (in the beginning, C_A inside the cylinder is zero because it contains just air), element A will flow from the region with a higher concentration of A (the top of the cylinder) to the region with a lower concentration (inside the cylinder). Let us consider that, by some mechanism that is chemically possible, the concentration of A at the bottom of the cylinder vase is

Fig. 4.14 Cylinder vase
initially containing air:
diffusion of A from the top
to the bottom

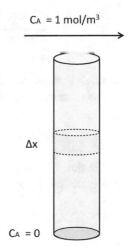

$C_A = 1\ \text{mol/m}^3$

Δx

$C_A = 0$

always zero. What would be the axial profile of the concentration of A inside the
cylinder vase over time?

Solution: The first thing to do to solve this problem is to define the control volume,
which, in this case, is a small cylinder with radius R and length Δx, as depicted in
Fig. 4.14. Since we want to know the axial profile of the concentration of A inside
the cylinder vase over time, the accumulation term must be considered. There is
neither generation nor consumption of A along the cylinder, so the conservation law
applied to the control volume yields $E - L = A$.

The flow of A from x to $x + \Delta x$ occurs by diffusion, due to the difference in the
concentration of A. The mass flow by diffusion, according to Fick's law is (see
Eq. 4.11):

$$(J_A)_x = -DA\left(\frac{dC_A}{dx}\right)$$

So, the amount of A at x and at $x + \Delta x$ in the control volume is shown as follows:

	x	$x + \Delta x$
Mass flow by diffusion (mol/s)	$-D\pi R^2\left(\dfrac{dC_A}{dx}\right)$	$-D\pi R^2\left(\dfrac{dC_A}{dx}\right) + \dfrac{d}{dx}\left(-D\pi R^2\left(\dfrac{dC_A}{dx}\right)\right)\Delta x$

in which (using international units):

$D =$ diffusivity of component A in the air (m^2/s)
$\pi R^2 =$ area from where the diffusive transport of A takes place (m^2)
$dC_A/dx =$ concentration gradient for component A $(\text{mol/m}^3\ \text{m})$

The accumulation of A (in mol) inside the control volume (V) in a period
Δt (s) can be obtained by the difference between the amounts of A at times $t + \Delta t$
and t.

	t	$t + \Delta t$
Amount of A (mol)	$V C_A$	$V C_A + \dfrac{d(V C_A)}{dt} \Delta t$

in which $V = \pi R^2 \Delta x = $ control volume (m^3).

The accumulation of A at time Δt can be written as:

$$\text{Accumulation of A (mol)} = V \frac{dC_A}{dt} \Delta t = \pi R^2 \Delta x \frac{dC_A}{dt} \Delta t$$

The flow of A is given in mol/s and the accumulation term is in mol, so the flow of A is multiplied by Δt to yield the mass balance shown in Eq. (4.22):

$$-D\pi R^2 \frac{\partial C_A}{\partial x} \Delta t - \left[-D\pi R^2 \frac{\partial C_A}{\partial x} + \frac{\partial}{\partial x}\left(-D\pi R^2 \frac{\partial C_A}{\partial x}\right) \Delta x \right] \Delta t = \pi R^2 \Delta x \frac{\partial C_A}{\partial t} \Delta t$$

$$(4.22)$$

Simplifying terms and assuming that the diffusion coefficient is steady, one can obtain:

$$\frac{\partial C_A}{\partial t} = D \frac{\partial^2 C_A}{\partial x^2} \tag{4.23}$$

The initial and boundary conditions needed to solve this PDE are:

At $t = 0$ s, $C_A = 0$ mol/m^3, for $0 \leq x \leq 0.05$ m
At $x = 0$ m, $C_A = 0$ mol/m^3, for $t > 0$ s
At $x = 0.05$ m, $C_A = 1$ mol/m^3, for $t > 0$ s

Equation (4.23) can be solved to generate the axial profiles of the concentration of A over time with the same shapes of the ones shown in Fig. 4.7b. The higher the diffusion coefficient, the faster a steady state is reached.

If the axial profile of the concentration of A inside the cylinder vase is needed only in a steady state, the mass balance would yield:

$$\frac{d^2 C_A}{dx^2} = 0 \tag{4.24}$$

with the boundary conditions:

At $x = 0$ m, $C_A = 0$ mol/m^3
At $x = 0.05$ m, $C_A = 1$ mol/m^3

Equation (4.24) can be integrated twice to generate a straight line, as foreseen by Fig. 4.7b. Example 4.7 can be solved starting the x-axis both at the bottom or at the top of the cylinder vase.

Until now, all examples presented in Chap. 4 have considered variations in an axial direction. Example 4.8 will assume changes of temperature in a radial

Fig. 4.15 Two concentric
cylinders with radial heat
conduction along the
aluminum annulus

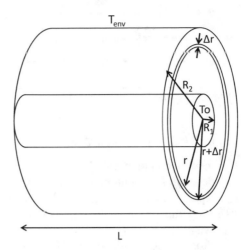

direction. The same procedure presented so far for developing models can be used.
However, it is interesting to pay attention to the control volume and to the area
considered for the flow of energy by conduction.

Example 4.8 Consider a solid cylinder of copper of length L and radius R_1.
Assume that the temperature of this cylinder remains constant and equal to T_0.
The copper cylinder is coated with an annulus made of aluminum, initially at
temperature T_1. The total radius of the concentric cylinders (copper plus aluminum)
is R_2, as depicted in Fig. 4.15. The environmental temperature is constant and equal
to T_{env}. Assume that $T_0 > T_1 > T_{env}$. Although the aluminum annulus exchanges
heat with the environment, the two ends of the two concentric cylinders are
insulated. Develop a mathematical model able to predict the radial profiles of the
temperature in the aluminum annulus over time from the beginning (when $T = T_1$)
until a steady state is reached.

Solution: Initially the annulus is at temperature T_1, but its temperature starts
changing radially, becoming higher near the copper cylinder and lower close to
the environment. There is no reason for axial changes in temperature, because there
is no driving force in this direction (the ends are insulated).

The first thing to do in order to model this system is to define a control volume.
In our case, we can consider a control volume with a small thickness Δr and length
L, as depicted in Fig. 4.15.

The conservation law applied to this control volume yields E $-$ L $=$ A.

As the temperature of the copper cylinder is higher, heat in the aluminum
annulus will flow from inside to outside. The energy that enters (at r) and leaves
(at $r + \Delta r$) the control volume is due to the conduction of heat (molecule to
molecule). These two terms plus the accumulation of energy in the control volume
can be written as follows:

	r (enters)	$r + \Delta r$ (leaves)
Heat flow rate by conduction (J/s)	$-kA\dfrac{dT}{dr}$	$-kA\dfrac{dT}{dr} + \dfrac{d}{dr}\left(-kA\dfrac{dT}{dr}\right)\Delta r$

	t	$t + \Delta t$
Amount of energy in the control volume (J)	$\rho V c_p T$	$\rho V c_p T + \dfrac{d(\rho V c_p T)}{dt}\Delta t$

in which (using international units):

k = thermal conductivity of the aluminum (J/s m °C)
ρ = density of the aluminum (kg/m^3)
c_p = specific heat of the aluminum (J/kg °C)
dT/dr = temperature gradient in a radial direction (°C/m)
T = temperature (°C)
A = cross-sectional area from where the heat flows by conduction (m^2)
V = control volume (m^3)

At this point it is very important to understand how the cross-sectional area (A) and the control volume (V) are calculated.

The area that the energy "sees" when flowing radially by conduction is $2\pi r L$, which is the superficial area of the aluminum annulus: the perimeter of the ring ($2\pi r$) times its length (L).

The volume of the control volume is $2\pi r \Delta r L$ and is calculated considering its area of the base ($2\pi r \Delta r$) times its length (L). You can visualize the area of the base, imagining the thin ring of the control volume being cut and stretched, making it similar to a rectangle with sides $2\pi r$ and Δr, yielding an area equal to $2\pi r \Delta r$. Alternativelly, you can calculate the area of the base of the ring as $\pi(r + \Delta r)^2 - \pi r^2$, what would yield $2\pi r \Delta r + \pi(\Delta r)^2$. Since Δr is very small, the square of Δr is much smaller than Δr, so $\pi(\Delta r)^2$ can be neglected, and you would obtain the same result for the area of the base ($2\pi r \Delta r$).

Now we can write the energy balance, multiplying the heat flow by Δt, to make the units compatible:

$$\underbrace{-k2\pi r L\frac{dT}{dr}\Delta t}_{\text{Enters}} - \underbrace{\left[-k2\pi r L\frac{dT}{dr} + \frac{d}{dr}\left(-k2\pi r L\frac{dT}{dr}\right)\Delta r\right]\Delta t}_{\text{Leaves}} = \underbrace{\frac{d(\rho 2\pi r \Delta r L c_p T)}{dt}\Delta t}_{\text{Accumulates}}$$

(4.25)

Simplifying terms, the energy balance can be rewritten as:

$$k\frac{\partial}{\partial r}\left(r\frac{\partial T}{\partial r}\right) = r\rho c_p \frac{\partial T}{\partial t}$$

(4.26)

or:

$$\frac{\partial^2 T}{\partial r^2} + \frac{1}{r}\frac{\partial T}{\partial r} = \frac{\rho c_p}{k}\frac{\partial T}{\partial t} \tag{4.27}$$

Observe that the terms 2, π, L, k, c_p, and ρ are taken outside the derivative because they are constant and do not depend on either the time or the radius (we assume that k, c_p, and ρ do not vary with temperature). However, the expression for the cross-sectional area $(2\pi r L)$ presents the independent variable r (radius) that must be held inside the partial derivative with respect to r (see Eq. 4.26).

The Eq. (4.27) needs two boundary conditions (related to the radius) and one initial condition (at $t = 0$). The conditions at $t = 0$ and $r = R_1$ are easily obtained, as follows:

At $t = 0$, $T = T_1$ for $R_1 \leq r \leq R_2$
At $r = R_1$ (for $t > 0$), $T = T_0$ (at this point, the annulus of aluminum is in contact with the cylinder of copper, which is kept at T_0)

The heat flows by conduction inside the aluminum annulus until it reaches the end at $r = R_2$. At this point $(r = R_2)$, the aluminum annulus exchanges heat with the environment, which is at T_{env}. In this way, we can say that at $r = R_2$, the heat that gets to R_2 by conduction is equal to the heat that leaves the annulus $(r = R_2)$ by convection (exchanging heat with the environment), or:

$$-kA\frac{dT}{dr} = hA'(T - T_{env}) \tag{4.28}$$

The area is the same for both terms of Eq. (4.28) ($A = A' = 2\pi R_2 L$) and can be simplified. So the three conditions needed to solve Eq. (4.27) are:

$$At\ t = 0,\quad T = T_1,\quad for\ R_1 \leq r \leq R_2$$

$$At\ r = R_1,\quad T = T_0,\quad for\ t > 0$$

$$At\ r = R_2,\quad \frac{dT}{dr} = -\frac{h}{k}(T - T_{env}),\quad for\ t > 0$$

Observe that the convection term does not appear in the energy balance Eqs. (4.26) and (4.27), only in the boundary conditions. This is because the control volume being considered (see Fig. 4.15) does not interface with the environment. This only occurs at $r = R_2$. Therefore, the convection term appears only in the boundary condition. A different situation occurred in Examples 4.1, 4.2, and 4.6, in which the control volume interfaced with the environment, so the convection term appeared in the energy balance equation.

The next example will present a more comprehensive situation, in which chemical reactions also occur.

Example 4.9 Consider a tubular reactor, shown in Fig. 4.16, in which an irreversible and exothermic reaction A $\overset{k}{\rightarrow}$ B takes place. This reactor, with

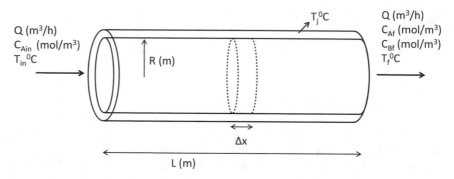

Fig. 4.16 Tubular reactor producing component B

radius R and length L, has a jacket at a constant temperature T_j °C, needed to control the reactor temperature. The reactant A is fed into the reactor at a flow rate Q m³/h, concentration $C_{A_{in}}$, and temperature T_{in} °C. The concentrations of A and B and the temperature of the fluid that leaves the reactor are C_{A_f}, C_{B_f}, and T_f, respectively. A chemical engineer needs to build another tubular reactor to produce B in a different plant. To save money, the engineer wants to know if a shorter reactor could be used to produce the same amount of B. There is a suspicion that most of A is consumed well before the end of the reactor. Develop a mathematical model to simulate this reactor in order to check the viability of building a shorter tube. Assume that the system operates in a steady state and there are no radial or angular profiles of the concentration and temperature inside the reactor.

Solution: The first thing to do in order to model this system is to define the control volume. In this problem, it is assumed there are no radial and angular profiles of the concentration and temperature, so all variations occur only along the length. In this way, the control volume will be a small slice of the reactor, with length equal to Δx, as represented by Fig. 4.16.

If there were variations in concentration and temperature also along the radius, the control volume would be different. We will revisit this problem in Example 4.11, when we will consider variations in more than one dimension.

The next thing to do to model this reactor is to apply the conservation law to the control volume, to obtain the mass and energy balances for the reactor. The jacket temperature remains invariable, so there is no need to develop an energy balance for the jacket. The conservation law applied to the control volume, keeping in mind that the reactor operates in a steady state, yields:

Mass Balance for the reactant A $\Rightarrow$ E − L − C = 0
Mass Balance for the product B $\Rightarrow$ E − L + G = 0
Energy Balance for the reactor $\Rightarrow$ E − L + G/C = 0

The amount of A that enters the control volume (at x) depends on the flow rate Q (m^3/h) and on the concentration C_A (mol/m^3), so the amount of A that goes into the control volume due to the fluid movement is QC_A (mol/h). This is the most important contribution to A entering the control volume. However, for systems in which the flow rate is very low and/or for viscous fluids, the diffusion contribution may be important, so in this problem, we will also consider the axial mass and heat diffusion. Table 4.2 shows the E (entering) and L (leaving) terms for the mass and energy balance, in which:

k = thermal conductivity of the fluid inside the reactor (kJ/h m $^{\circ}$C)
ρ = density of the fluid inside the reactor (kg/m^3)
c_p = specific heat of the fluid inside the reactor (kJ/kg $^{\circ}$C)
D = diffusivity of reactant A or product B in the fluid inside the reactor (m^2/h)
πR^2 = area from where the diffusive transport of mass and energy takes place (m^2)
Q = volumetric flow rate inside the reactor (m^3/h)

We will assume that the diffusion coefficient D is the same for reactant A and product B, and that its value remains invariable. We will also consider that the fluid that travels through the reactor remains invariable values for density (ρ), specific heat (c_p), and thermal conductivity (k).

Besides the terms shown in Table 4.2, the heat exchanged by convection between the reaction mixture and the jacket must be added to the energy balance.

Heat transfer by convection : $U(2\pi R\Delta x)(T_j - T)$ (kJ/h)

Observe that the exchange area considered in the transfer by convection ($2\pi R\Delta x$) is different from the cross-sectional area (πR^2) used in Table 4.2. Do not confuse the

Table 4.2 Amounts of A, B, and energy entering and leaving the control volume

	x (enters)	$x + \Delta x$ (leaves)
Flow of A due to fluid movement (mol/h)	QC_A	$QC_A + \dfrac{d(QC_A)}{dx}\Delta x$
Flow of A due to diffusion (mol/h)	$-D\pi R^2\dfrac{dC_A}{dx}$	$-D\pi R^2\dfrac{dC_A}{dx} + \dfrac{d}{dx}\left(-D\pi R^2\dfrac{dC_A}{dx}\right)\Delta x$
Flow of B due to fluid movement (mol/h)	QC_B	$QC_B + \dfrac{d(QC_B)}{dx}\Delta x$
Flow of B due to diffusion (mol/h)	$-D\pi R^2\dfrac{dC_B}{dx}$	$-D\pi R^2\dfrac{dC_B}{dx} + \dfrac{d}{dx}\left(-D\pi R^2\dfrac{dC_B}{dx}\right)\Delta x$
Flow of energy due to fluid movement (J/h)	$Q\rho c_p T$	$Q\rho c_p T + \dfrac{d(Q\rho c_p T)}{dx}\Delta x$
Flow of energy due to conduction (J/h)	$-k\pi R^2\dfrac{dT}{dx}$	$-k\pi R^2\dfrac{dT}{dx} + \dfrac{d}{dx}\left(-k\pi R^2\dfrac{dT}{dx}\right)\Delta x$

areas. Also observe that there is no mass transfer through the jacket, so no term needs to be added to the mass balances related to that.

The last terms missing in the mass and energy balances are the consumption and generation terms, written as follows:

$$\text{Rate of reaction (kmol/h)} \Rightarrow k'C_A V$$

$$\text{Heat of reaction (kJ/h)} \quad \Rightarrow k'C_A V(-\Delta H_R)$$

in which:

$k' = $ constant of reaction (1/h)
$V = \pi R^2 \Delta x = $ control volume (m^3)
$\Delta H_R = $ heat of reaction (kJ/kmol)

Considering all terms in Table 4.2, plus the terms related to heat convection and reaction, and afterwards simplifying terms, the mass and energy balances can be written as follows:

$$-Q\frac{dC_A}{dx} + D\,\pi R^2\frac{d^2C_A}{dx^2} - k'\pi R^2 C_A = 0 \qquad (4.29)$$

$$-Q\frac{dC_B}{dx} + D\,\pi R^2\frac{d^2C_B}{dx^2} + k'\pi R^2 C_A = 0 \qquad (4.30)$$

$$-Q\rho c_p\frac{dT}{dx} + k\pi R^2\frac{d^2T}{dx^2} + U2\pi R(T_j - T) + k'\pi R^2 C_A(-\Delta H_R) = 0 \qquad (4.31)$$

Observe that the Arrhenius constant k' depends on T and that the energy balance depends on C_A, so the mass and energy equations must be solved simultaneously.

To solve this system with three ODEs, two boundary conditions (for each equation) are needed. In our problem, we know the concentrations and temperature at $x = 0$:

At $x = 0$: $C_A = C_{A_{in}}$; $C_B = C_{B_{in}} = 0$; $T = T_{in}$

We need another condition to solve the equation system. We can use one of the following conditions:

At $x = L$: $C_A = C_{A_f}$; $C_B = C_{B_f}$; $T = T_f$ or
At $x \to \infty$: $C_A = 0$; $C_B = C_{A_{in}}$; $T = T_j$ and $dC_A/dx = dC_B/dx = dT/dx = 0$

Observe that if the reactor is very long, all of the reactant will be consumed and the reaction mixture will exchange heat with the jacket until its temperature reaches T_j. Also observe that the gradient of the concentration reduces along the length as the reactant is consumed, and reaches zero when there is no reactant available. The same behavior is observed for the gradient of the temperature when the reactant finishes and the reactor temperature reaches T_j.

If the diffusion terms were not considered in the mass and energy balance, the model equations for this reactor would become (compare this with Eqs. 4.29, 4.30 and 4.31):

$$-Q\frac{dC_A}{dx} - k'\pi R^2 C_A = 0 \tag{4.32}$$

$$-Q\frac{dC_B}{dx} + k'\pi R^2 C_A = 0 \tag{4.33}$$

$$-Q\rho c_p \frac{dT}{dx} + U2\pi R(T_j - T) + k'\pi R^2 C_A(-\Delta H_R) = 0 \tag{4.34}$$

That is solved using the boundary conditions:

At $x = 0$: $C_A = C_{A_{in}}$; $C_B = C_{B_{in}} = 0$; $T = T_{in}$

The concentration and temperature profiles obtained from solution of Eqs. (4.29), (4.30) and (4.31) (considering diffusional effects) or Eqs. (4.32), (4.33) and (4.34) (not considering diffusional effects) can be very similar. This occurs because the contributions due to the flow rate (QC_A, QC_B and $Q\rho c_p T$) are much more important than the diffusion contributions ($-D\pi R^2 \frac{dC_A}{dx}$, $-D\pi R^2 \frac{dC_B}{dx}$ and $-k\pi R^2 \frac{dT}{dx}$). The diffusion effects should be considered when an accurate model is needed to simulate systems where the flow rate is low and/or the fluid viscosity is high.

The Sect. 4.3 will consider problems where variations occur in more than one spatial dimension. In this next section, it is important to pay attention to how the control volume is defined.

4.3 Examples Considering Variations in More than One Dimension

In this section, more comprehensive examples will be presented, with variations in more than one dimension. Let us start with an example considering the axial and radial heat transfer by conduction in a cylindrical bar. This example is a combination of the concepts presented previously in Examples 4.6 and 4.8.

Example 4.10 Imagine a cylindrical bar made of a material with low thermal conductivity. This bar, with length (L) of 1 m and radius (R) equal to 0.3 m, is initially at 50 °C (Fig. 4.17). At some point, one of the two ends of the bar is fixed to a wall that is at 70 °C (assume that the temperature of the wall does not change with time). Assume that the bar exchanges heat with the environment, which is at 25 °C. One wants to know the radial and axial profiles of the temperature after a steady state is reached. Assume there is no angular profile of the temperature inside the bar.

Solution: One can imagine that the cylinder is at 50 °C initially, but its temperature close to the wall starts increasing when it is fixed to the wall, generating an axial profile of the temperature. At the same time, the cylinder starts exchanging heat

Fig. 4.17 Cylindrical bar
with heat transfer in axial
and radial directions

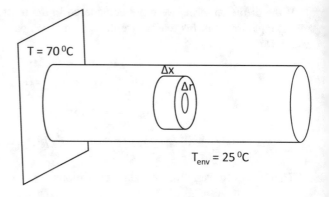

Table 4.3 Energy entering and leaving the control volume in an axial direction

	x (enters)	$x + \Delta x$ (leaves)
Axial flow of energy by conduction (J/h)	$-k2\pi r\Delta r\dfrac{dT}{dx}$	$-k2\pi r\Delta r\dfrac{dT}{dx} + \dfrac{d}{dx}\left(-k2\pi r\Delta r\dfrac{dT}{dx}\right)\Delta x$

with the environment, generating also a radial profile of the temperature (with a higher temperature in the center). Since the radius is big and the thermal conductivity of the material is low, the radial profile of the temperature can be significant and must be considered.

To model this system, first we need to define a control volume small enough to guarantee the same temperature inside it. Since the temperature changes along x and r, we need to consider a small ring with length Δx and thickness Δr as the control volume, as shown in Fig. 4.17. The control volume for this problem can be visualized as an intersection of the control volumes drawn in Figs. 4.10 and 4.15.

Since one wants to know the temperature profiles after a steady state is reached, and since there is no chemical reaction, the conservation law applied to the control volume shown in Fig. 4.17 yields:

$$E - L = 0$$

The heat enters the control volume at x and r, and leaves it at $x + \Delta x$ and $r + \Delta r$. The heat flow inside the control volume is only by conduction. The convection occurs only at $r = R = 0.3$ m (for all x) and at $x = L = 1$ m (for all r), which are regions outside the control volume.

Now, using the concept of the infinitesimal variation of the dependent variable with an independent variable, Tables 4.3 and 4.4 can be created.

Observe that the cross-sectional area from where the energy goes in a radial direction is $2\pi r\Delta x$, which is similar to the area found in Example 4.8, except for the length, which was L and now is Δx. The cross-sectional area from where the energy passes axially is $2\pi r\Delta r$, which is the area of the base of the control volume, also calculated in Example 4.8.

Table 4.4 Energy entering and leaving the control volume in a radial direction

	r (enters)	$r + \Delta r$ (leaves)
Radial flow of energy by conduction (J/h)	$-k2\pi r\Delta x\dfrac{dT}{dr}$	$-k2\pi r\Delta x\dfrac{dT}{dr} + \dfrac{d}{dr}\left(-k2\pi r\Delta x\dfrac{dT}{dr}\right)\Delta r$

Considering the thermal conductivity constant along the cylindrical bar, we can use the terms in Tables 4.3 and 4.4 to obtain the energy balance. After simplifying terms, the energy balance becomes:

$$r\frac{\partial^2 T}{\partial x^2} + \frac{\partial}{\partial r}\left(r\frac{\partial T}{\partial r}\right) = 0 \tag{4.35}$$

This PDE can be solved using four boundary conditions (at $x = 0$, $r = 0$, $x = L$, and $r = R$). It is known that the temperature at $x = 0$ is equal to the temperature of the wall (70 °C). At $x = L$, the heat that comes from conduction goes out by convection. The same thing occurs at $r = R$. One can imagine there is a symmetrical radial profile of the temperature inside the cylinder, with a maximum temperature in the center. This symmetry condition gives us the fourth condition needed to solve the PDE: at $r = 0$, $dT/dr = 0$. So we can write the following boundary conditions:

At $x = 0$: $T = 70\,^{\circ}\mathrm{C}$ (4.36)

At $x = L$: $-k2\pi r\Delta r\dfrac{dT}{dx} - h2\pi r\Delta r(T - T_{\mathrm{env}})$ or $\dfrac{dT}{dx} = -\dfrac{h}{k}(T - T_{\mathrm{env}})$ (4.37)

At $r = 0$: $\dfrac{dT}{dr} = 0$ (4.38)

At $r = R$: $-k2\pi r\Delta x\dfrac{dT}{dr} = h2\pi r\Delta x(T - T_{\mathrm{env}})$ or $\dfrac{dT}{dr} = -\dfrac{h}{k}(T - T_{\mathrm{env}})$ (4.39)

Observe that despite the exchange of heat by convection between the cylinder and the environment, this convection term does not appear in the energy balance, only in the boundary conditions. This is because the control volume does not interface with the environment.

If one wants to know the axial and radial profiles of the temperature in the transient regime, the accumulation term shown in Table 4.5 must be added to the conservation law.

Keeping in mind that the control volume is $2\pi r\Delta r\Delta x$, the PDE that represents this system in the transient regime will be:

$$\frac{\rho c_p}{k}\frac{\partial T}{\partial t} = \frac{\partial^2 T}{\partial x^2} + \frac{1}{r}\frac{\partial}{\partial r}\left(r\frac{\partial T}{\partial r}\right) \tag{4.40}$$

Besides boundary conditions (4.36)–(4.39), the initial condition needed to solve Eq. (4.40) is:

At $t = 0$, $T = 50$ °C for the entire bar

Table 4.5 Calculus of the accumulation term for the cylindrical bar

	t	$t + \Delta t$
Amount of energy in the control volume (J)	$V\rho c_p T$	$V\rho c_p T + \dfrac{d(V\rho c_p T)}{dt}\Delta t$

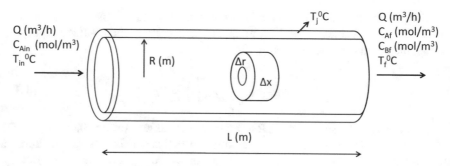

Fig. 4.18 Tubular reactor with changes in axial and radial directions

Example 4.11 This next example will revisit Example 4.9, but now we will assume that the reactor exhibits also radial profiles of the concentration and temperature. The tubular reactor, where the reaction A $\xrightarrow{k}$ B takes place, has a jacket that is at a constant temperature of T_j °C. All operating conditions and properties are the same of the as those in Example 4.9. One wants to know the axial and radial profiles of the temperature and concentration inside this reactor.

Solution: The main difference between the solution presented in Example 4.9 and the one presented herein is the definition of the control volume. Since we are considering that both axial and radial variations are important, we need to consider a small ring with length Δx and thickness Δr to represent the control volume, as shown in Fig. 4.18 (compare this with Fig. 4.16).

The conservation law applied to this example generates the same expressions presented previously (rewritten below). Remember that we are considering a steady state (the accumulation term is equal to zero).

$$
\begin{aligned}
\text{Mass Balance for the reactant A} &\Rightarrow E - L - C = 0 \\
\text{Mass Balance for the product B} &\Rightarrow E - L + G = 0 \\
\text{Energy Balance for the reactor} &\Rightarrow E - L + G/C = 0
\end{aligned}
$$

The amounts of material and energy that enter and leave the control volume can be obtained using the concept of infinitesimal variation of the dependent variable with the independent variable.

As was done in Example 4.9, we will also consider the diffusion of mass and energy. The amounts of reactant A, product B, and energy that enter and leave the control volume in the direction x can be seen in Table 4.6.

Observe that Table 4.6 is very similar to Table 4.2; however, the terms are adapted in order to consider the new control volume.

Table 4.6 A, B, and energy entering and leaving the control volume in an axial direction

	x (enters)	$x + \Delta x$ (leaves)
Flow of A due to fluid movement (mol/h)	$Q\left(\dfrac{2\pi r\Delta r}{\pi R^2}\right)C_A$	$Q\left(\dfrac{2\pi r\Delta r}{\pi R^2}\right)C_A + \dfrac{d}{dx}\left[Q\left(\dfrac{2\pi r\Delta r}{\pi R^2}\right)C_A\right]\Delta x$
Flow of A due to diffusion (mol/h)	$-D2\pi r\Delta r\dfrac{dC_A}{dx}$	$-D2\pi r\Delta r\dfrac{dC_A}{dx} + \dfrac{d}{dx}\left(-D2\pi r\Delta r\dfrac{dC_A}{dx}\right)\Delta x$
Flow of B due to fluid movement (mol/h)	$Q\left(\dfrac{2\pi r\Delta r}{\pi R^2}\right)C_B$	$Q\left(\dfrac{2\pi r\Delta r}{\pi R^2}\right)C_B + \dfrac{d}{dx}\left[Q\left(\dfrac{2\pi r\Delta r}{\pi R^2}\right)C_B\right]\Delta x$
Flow of B due to diffusion (mol/h)	$-D2\pi r\Delta r\dfrac{dC_B}{dx}$	$-D2\pi r\Delta r\dfrac{dC_B}{dx} + \dfrac{d}{dx}\left(-D2\pi r\Delta r\dfrac{dC_B}{dx}\right)\Delta x$
Flow of energy due to fluid movement (J/h)	$Q\left(\dfrac{2\pi r\Delta r}{\pi R^2}\right)\rho c_p T$	$Q\left(\dfrac{2\pi r\Delta r}{\pi R^2}\right)\rho c_p T + \dfrac{d}{dx}\left[Q\left(\dfrac{2\pi r\Delta r}{\pi R^2}\right)\rho c_p T\right]\Delta x$
Flow of energy due to conduction (J/h)	$-k2\pi r\Delta r\dfrac{dT}{dx}$	$-k2\pi r\Delta r\dfrac{dT}{dx} + \dfrac{d}{dx}\left(-k2\pi r\Delta r\dfrac{dT}{dx}\right)\Delta x$

Table 4.7 A, B, and energy entering and leaving the control volume in a radial direction

	r (enters)	$r + \Delta r$ (leaves)
Flow of A due to diffusion (mol/h)	$-D2\pi r\Delta x\dfrac{dC_A}{dr}$	$-D2\pi r\Delta x\dfrac{dC_A}{dr} + \dfrac{d}{dr}\left(D2\pi r\Delta x\dfrac{dC_A}{dr}\right)\Delta r$
Flow of B due to diffusion (mol/h)	$-D2\pi r\Delta x\dfrac{dC_B}{dr}$	$D2\pi r\Delta x\dfrac{dC_B}{dr} + \dfrac{d}{dr}\left(-D2\pi r\Delta x\dfrac{dC_B}{dr}\right)\Delta r$
Flow of energy by conduction (J/h)	$-k2\pi r\Delta x\dfrac{dT}{dr}$	$-k2\pi r\Delta x\dfrac{dT}{dr} + \dfrac{d}{dr}\left(-k2\pi r\Delta x\dfrac{dT}{dr}\right)\Delta r$

The cross-sectional area for the diffusion terms in the x direction change from πR^2 to $2\pi r\Delta r$. The flow rate Q is related to the *total* cross-sectional area (πR^2). However, the amount of fluid crossing in an axial direction the new cross-sectional area ($2\pi r\Delta r$) is much smaller and is proportional to this new area. In this way, the flow rate crossing the new control volume axially is $Q\left(\frac{2\pi r\Delta r}{\pi R^2}\right)$.

The amounts of reactant A, product B, and energy that enter and leave the control volume in a radial direction can be seen in Table 4.7. The fluid is fed only in an axial direction, so the contribution due to the fluid movement in a radial direction is zero. The differences in the concentration and temperature along the radius are due to diffusion of mass and heat conduction. Observe that in Table 4.7, the cross-sectional area for the radial diffusion is $2\pi r\Delta x$ (the same used in Table 4.4). Also observe that the heat transfer by convection is not considered in the conservation law and will appear only in the boundary conditions, because the control volume does not interface with the jacket.

The generation and consumption terms are represented in the same way as was done in Example 4.9; however, the volume is calculated differently.

$$\text{Rate of reaction (mol/h)} \Rightarrow k' C_A V$$

$$\text{Heat of reaction (J/h)} \Rightarrow k' C_A V(-\Delta H_R)$$

in which $V = 2\pi r \Delta r \Delta x$ = control volume (m³).

All terms in Tables 4.6 and 4.7 plus the reaction terms are combined to generate the system of equations that represent this reactor, as follows:

Balance for reactant A:

$$-\frac{Q}{\pi R^2} \frac{\partial C_A}{\partial x} + D\frac{\partial^2 C_A}{\partial x^2} + \frac{D}{r}\frac{\partial}{\partial r}\left(r\frac{\partial C_A}{\partial r}\right) - k' C_A = 0 \qquad (4.41)$$

Balance for product B:

$$-\frac{Q}{\pi R^2} \frac{\partial C_B}{\partial x} + D\frac{\partial^2 C_B}{\partial x^2} + \frac{D}{r}\frac{\partial}{\partial r}\left(r\frac{\partial C_B}{\partial r}\right) + k' C_A = 0 \qquad (4.42)$$

Balance of energy:

$$-\frac{Q\rho c_p}{\pi R^2} \frac{\partial T}{\partial x} + k\frac{\partial^2 T}{\partial x^2} + \frac{k}{r}\frac{\partial}{\partial r}\left(r\frac{\partial T}{\partial r}\right) + k' C_A(-\Delta H_R) = 0 \qquad (4.43)$$

Observe that the terms Δx and Δr can be simplified. It is assumed that Q, D, ρ, c_p, and k are constant and can be removed from the derivative. Note that $Q/\pi R^2$, present in Eqs. (4.41), (4.42) and (4.43), is the fluid velocity.

To solve this equation system, we need four boundary conditions: two in the x- and two in the r-axes. The two conditions in x (at $x = 0$ and $x = L$) are the same as those presented in Example 4.9. As occurred in Examples 4.8 and 4.10, the boundary condition at $r = 0$ is obtained due to the radial symmetry. The other condition is given in $r = R$. For the energy balance, we can assume that the heat that reaches $r = R$ by conduction leaves the tube by convection, exchanging heat with the jacket. This provides us with a boundary condition at $r = R$, as per Examples 4.8 and 4.10. For the mass balances, the boundary condition at $r = R$ is based on the fact that the tube wall is nonpermeable for mass transport, so the radial gradients of the concentration at $r = R$ are equal to zero. So the set of conditions used to solve the system of PDEs is:

At $x = 0$, for $0 \le r \le R$: $\qquad C_A = C_{A_{in}}; C_B = C_{B_{in}} = 0; T = T_{in}$ $\qquad$ (4.44)

At $r = 0$, for $0 \le x \le L$: $\qquad \dfrac{dC_A}{dr} = \dfrac{dC_B}{dr} = \dfrac{dT}{dr} = 0$ $\qquad$ (4.45)

At $x = L$, for $0 \le r \le R$: $\qquad C_A = C_{A_f}; C_B = C_{B_f} = 0; T = T_f$ $\qquad$ (4.46)

At $r = R$, for $0 \le x \le L$: $\qquad \dfrac{dC_A}{dr} = 0; \quad \dfrac{dC_B}{dr} = 0; \quad \dfrac{dT}{dr} = -\dfrac{h}{k}(T - T_j)$ $\qquad$ (4.47)

Table 4.8 Calculus of accumulation terms for the tubular reactor

	t	$t + \Delta t$
Amount of A (mol)	$2\pi r \Delta r \Delta x C_A$	$2\pi r \Delta r \Delta x C_A + \dfrac{d}{dt}(2\pi r \Delta r \Delta x C_A)\Delta t$
Amount of B (mol)	$2\pi r \Delta r \Delta x C_B$	$2\pi r \Delta r \Delta x C_B + \dfrac{d}{dt}(2\pi r \Delta r \Delta x C_B)\Delta t$
Amount of energy in the control volume (J)	$2\pi r \Delta r \Delta x \rho c_p T$	$2\pi r \Delta r \Delta x \rho c_p T + \dfrac{d}{dt}(2\pi r \Delta r \Delta x \rho c_p T)\Delta t$

Since there are radial profiles of the concentration and temperature along all of the reactor length, these radial profiles also exist at the exit of the reactor. The concentration and temperature values considered in the boundary condition $x = L$ could be an average of the values over the radius.

If a mathematical model to predict this reactor in the transient regime is also needed, the accumulation terms must be considered in the conservation law, as calculated in Table 4.8.

The model to represent the reactor in the transient regime can be represented by Eqs. (4.48), (4.49) and (4.50) (compare this with the model in a steady-state regime Eqs. (4.41), (4.42) and (4.43):

Balance for reactant A:

$$\frac{\partial C_A}{\partial t} = -\frac{Q}{\pi R^2}\frac{\partial C_A}{\partial x} + D\frac{\partial^2 C_A}{\partial x^2} + \frac{D}{r}\frac{\partial}{\partial r}\left(r\frac{\partial C_A}{\partial r}\right) - k'C_A \qquad (4.48)$$

Balance for product B:

$$\frac{\partial C_B}{\partial t} = -\frac{Q}{\pi R^2}\frac{\partial C_B}{\partial x} + D\frac{\partial^2 C_B}{\partial x^2} + \frac{D}{r}\frac{\partial}{\partial r}\left(r\frac{\partial C_B}{\partial r}\right) + k'C_A \qquad (4.49)$$

Balance of energy:

$$\rho c_p \frac{\partial T}{\partial t} = -\frac{Q\rho c_p}{\pi R^2}\frac{\partial T}{\partial x} + k\frac{\partial^2 T}{\partial x^2} + \frac{k}{r}\frac{\partial}{\partial r}\left(r\frac{\partial T}{\partial r}\right) + k'C_A(-\Delta H_R) \qquad (4.50)$$

The boundary conditions in r and x are the same as those presented previously (Eqs. (4.44), (4.45), (4.46) and (4.47)), and the initial condition needed to solve the transient model is:

$$\text{At } t = 0: \qquad C_A = C_{A_0}; \qquad C_B = C_{B_0}; \qquad T = T_0 \qquad (4.51)$$

The initial conditions at $t = 0$ (C_{A_0}, C_{B_0}, and T_0) can be a constant or can vary with x and/or r (functions of x and/or r).

Finding boundary conditions to solve differential equations generated from the modeling can be very tricky. Besides, depending on the boundary conditions used, numerical solution of the models can become less or more complicated.

Some works in the literature, such as Boyadjiev (2010), show examples applying boundary conditions in different ways.

Chapters 3 and 4 provide enough information to develop models for pieces of equipment used in chemical plants at different levels of complexity. Depending on the complexity of the system being modeled, different kinds of equations are generated. It is possible to obtain algebraic equations, ODEs (of first and second degrees), and PDEs (varying with 2, 3, or 4 independent variables). The second part of this book will focus on numerical solution of the equations that can be generated from the modeling. Chapters 5, 6, and 7 will solve systems represented by algebraic equations, ODEs, and PDEs, respectively, adopting Excel as a computational tool.

Proposed Problems

4.1) Imagine a very thin disc of thickness δ initially at 50 °C. This disc has a small orifice in the middle, in which a pin at a temperature of 100 °C is engaged. This disk exchanges heat with the environment, which is at 25 °C. Assume that both the environmental and pin temperatures do not change over time. Also assume there is no temperature profile along the thickness δ of the disc.

(a) Make a sketch of how the disc temperature would vary along the radius and over time until a steady state is reached. Draw curves on the same graph.
(b) Do the modeling of this system and find the PDE to represent how the temperature changes along the radius and over time.
(c) Define all initial/boundary conditions necessary to solve this PDE.

4.2) Consider the heat conduction on an extended surface represented by the fin shown in the Figure below, with dimensions $L \times L \times 10\,L$. The fin, which was initially at 25 °C, is fixed to a wall at 150 °C. The fin exchange heat with the environment, which is at 25 °C. Assume that the wall and environmental temperatures do not change over time.

(a) Find the PDE that represents the fin temperature along x, y, and z, and over time.
(b) Find the initial/boundary necessary to solve the PDE.

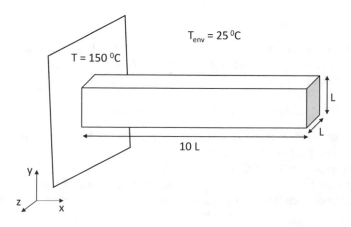

4.3) Imagine a solid cone, as depicted below. The lower and upper bases of the cone are 2 cm and 12 cm, respectively, and its height is 5 cm. This cone was initially at 35 °C, and the environmental temperature (T_{env}) is 25 °C. The lower base of the cone is submitted to a temperature T_b equal to 180 °C. The lateral of the cone is insulated, and the cone exchanges heat with the environment only through the upper base. Assume there are no radial and angular profiles of the temperature in the cone and that T_b and T_{env} do not change with time.

(a) Define the ODE that represents the variation of temperature along the height of the solid cone in a steady state. (Hint: represent the radius of the cone as a function of the height to obtain the ODE).
(b) Define the initial/boundary conditions needed to solve the ODE.

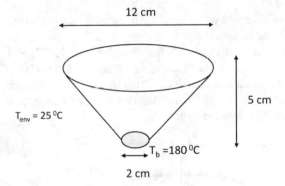

4.4) Imagine two tubular reactors with radius R_1 (m) and R_2 (m), set in a consecutive way, in which the exothermic reaction A → B takes place. This system exchanges heat with the environment, which is at T_{env} (°C). The radius of the first reactor is three times the radius of the second ($R_1 = 3R_2$). The lengths of the first and second reactors are L_1 and L_2 (m), respectively. Reactant A is fed into the reaction system at concentration C_{Ain} (mol/m³), temperature T_{in} (°C), and flow rate Q (m³/h), as depicted by the figure below. There is a sample collection between the two reactors, so it is possible to know the concentration and temperature of the reaction mixture at this point.

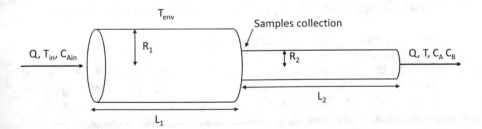

Assume that the reactors operate in a steady state. Consider that all properties of the fluid (density, specific heat, thermal conductivity, mass diffusivity) are the same for both reactors and do not vary along the length. The global coefficient of the heat transfer by convection for this reaction system is h (J/h m^2 °C).

(a) Draw the control volume and develop the energy and mass balance for reactant A in the first reactor. In this case, consider mass and energy diffusion only in an axial direction. Define all boundary conditions to solve the model equations.

(b) Draw the control volume and develop the energy and mass balance for reactant A in the second reactor. In this case, consider the mass and energy diffusion only in a radial direction. Define all boundary conditions to solve the model equations.

4.5) Imagine a solid sphere with radius R_s, hung by a very thin string in the center of a closed cylindrical recipient full of a fluid initially at 50 °C, as depicted in the figure below. The air that surrounds the cylinder is at 25 °C, and its temperature remains invariable over time. The solid sphere was at 100 °C before being immersed in the cylinder. It can be assumed that all properties of the fluid and of the sphere do not change over time. Assume that the temperature of the fluid is homogeneous along the entire cylinder.

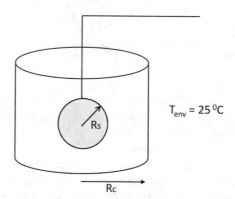

(a) Consider that the sphere is made of a very conductive material and that its radius R_s is very small; in a way, the temperature profiles inside the sphere are neglectable. Develop an energy balance for the sphere and for the fluid, and obtain the two differential equations that represent this system over time. Define the initial conditions.

(b) Now consider that the sphere is made of a material with very low thermal conductivity, and its radius R_s is big enough to make the radial temperature profiles inside it significant. Develop the energy balances for this new situation, and define all initial/boundary conditions. Is the energy balance for the fluid different in this case?

For cases (a) and (b), make assumptions for the models and define parameters if necessary. Remember that the area and volume of a sphere are $4\pi R^2$ and $(4/3)\pi R^3$, respectively.

4.6) Imagine a tubular reactor in which an irreversible reaction A $\rightarrow$ B takes place. Reactant A is fed into the reactor at a flow rate Q (m^3/h) and in a concentration ($C_{A_{in}}$) equal to 1 mol/L. The reactor has a jacket with a thermal fluid at T_j °C flowing in concurrently and exchanging heat with the reactional mixture and with the environment (at a constant temperature $T_{env} = 25$ °C). Assume there are no radial and angular profiles inside the reactor and jacket and that they operate in a steady state. Also assume that the diffusional effects in an axial direction are important. Find the differential equations system that represents the variation in the concentration of reactant A, the temperature of the reaction mixture, and the temperature of the jacket along the length. Find the boundary conditions needed to solve the equations system. Define the parameters and the hypotheses for your model when needed.

4.7) Imagine that a solid cylinder with a radius and length of 2 m and 10 m, respectively, initially at 30 °C, is fixed between two surfaces at constant temperatures (100 °C on the right and 20 °C on the left).

(a) Assuming there is no heat exchange with the environment (an insulated cylinder), do a sketch of the axial profiles of the temperature in the cylinder until a steady state is reached.

(b) Assuming that the thermal conductivity of the material is equal to 1 J/h m °C, what are the temperature and the energy flux at the positions $x = 2$ m, 5 m, and 9 m in a steady state?

References

Kern, Q.D.: Process Heat Transfer. McGraw-Hill Book Company, New York (1950)

Boyadjiev, C.: Theoretical Chemical Engineering Modeling and Simulation. Springer, Berlin (2010)

Bergman, T.L., Lavine, A.S., Incropera, F.P., Dewitt, D.P.: Fundamentals of Heat and Mass Transfer, 7th edn. Wiley, Hoboken (2011)

Bird, R.B., Steward, W.E., Lightfoot, E.N.: Transport Phenomena, 2nd edn. Wiley, New York (2007)

Fox, R.W., McDonald, A.T., Pritchard, P.J., Leylegian, J.C.: Fluid Mechanics, 8th edn. Wiley, Hoboken (2012)

Welty, J.R., Wicks, C.E., Wilson, R.E., Rorrer, G.L.: Fundamentals of Momentum, Heat and Mass Transfer, 5th edn. Wiley, Hoboken (2007)

Chapter 5
Solving an Algebraic Equations System

In this chapter, we will see a practical way to solve an algebraic equations system obtained from lumped-parameters models in a steady state. There are many different numerical methods to solve linear and nonlinear algebraic equations, but in this chapter just a few alternatives will be used, because the main objective of this book is to obtain a fast, robust, and simple way to simulate chemical engineering problems, not to study in detail the different numerical methods available in the literature. In order to make the problem solution even easier, all simulations will be done using Excel.

Sections 5.1 and 5.2 deal with problems involving linear and nonlinear algebraic equations systems, respectively. Only one numerical method for each section will be presented. Section 5.3 will present a third numerical approach, which can be used for both linear and nonlinear equations.

5.1 Problems Involving Linear Algebraic Equations

This chapter will provide only the essential information about matrixes needed for solution of algebraic equations. More information about this issue can be found in the specific literature (Isaacson and Keller 1966; Chapra and Canale 2005; Burden et al. 2014; etc).

The first information needed to understand this chapter is that a linear algebraic equations system can be represented as follows:

The original version of this chapter was revised. An erratum to this chapter can be found at https://doi.org/10.1007/978-3-319-66047-9_8

© Springer International Publishing AG 2018
L.M.F. Lona, *A Step by Step Approach to the Modeling of Chemical Engineering Processes*, https://doi.org/10.1007/978-3-319-66047-9_5

$$a_{11}x_1 + a_{12}x_2 + \cdots + a_{1n}x_n = b_1$$
$$a_{21}x_1 + a_{22}x_2 + \cdots + a_{2n}x_n = b_2$$
$$\cdots \qquad \cdots \qquad \cdots \quad \cdots$$
$$a_{n1}x_1 + a_{n2}x_2 + \cdots + a_{nn}x_n = b_n$$

In which a_{ij} and b_i are known numerical coefficients, and x_j are variables we want to calculate.

Or, in a matrix form:

$$[A]\{X\} = \{B\} \tag{5.1}$$

in which:

$$[A] = \begin{bmatrix} a_{11} & \cdots & a_{1n} \\ \vdots & \ddots & \vdots \\ a_{m1} & \cdots & a_{mn} \end{bmatrix}, \{X\} = \begin{Bmatrix} x_1 \\ \vdots \\ x_n \end{Bmatrix} \text{ and } \{B\} = \begin{Bmatrix} b_1 \\ \vdots \\ b_n \end{Bmatrix}$$

The next two concepts to remember are the definitions of the *identity matrix* and the *inverse of a matrix*.

The *identity matrix* $[I]$ is the matrix that plays, in the matrix algebra, the same role of the number "1" in the number system. It has the number 1 in the main diagonal and zeros elsewhere.

$$I = \begin{bmatrix} 1 & 0 & \cdots & 0 \\ 0 & 1 & \cdots & 0 \\ \vdots & \vdots & \ddots & \vdots \\ 0 & 0 & \cdots & 1 \end{bmatrix}$$

Given a matrix A, we can write:

$$[A][I] = [I][A] = [A] \tag{5.2}$$

The *inverse of a matrix* A is represented by $[A]^{-1}$ such that:

$$[A]\,[A]^{-1} = [A]^{-1}\,[A] = [I] \tag{5.3}$$

Keeping these two concepts in mind, the two sides of Eq. (5.1) can be multiplied by $[A]^{-1}$ to yield:

$$[A]^{-1}[A]\,\{X\} = [A]^{-1}\{B\} \tag{5.4}$$

Applying Eqs. (5.2) and (5.3) on the left side of Eq. (5.4) we obtain:

$$\{X\} = [A]^{-1}\{B\} \tag{5.5}$$

So in order to solve a linear algebraic equations system and find the vector $\{X\}$, the matrix $[A]$ (matrix of coefficients) must be inverted, and this inverted matrix must be multiplied by the vector B.

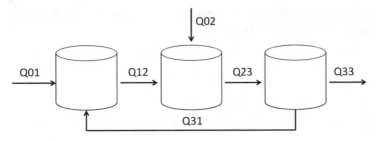

Fig. 5.1 Three interconnected tanks

Table 5.1 Volumetric flow rates for the three interconnected tanks

Q_{01}	Q_{02}	Q_{31}	$Q_{12}=Q_{01}+Q_{31}$	$Q_{23}=Q_{02}+Q_{12}$	$Q_{33}-Q_{23}-Q_{31}$	Units
5	1	2	7	8	6	m³/min

Table 5.2 Mass balance for the three tanks

Tank	Units (mol/min)
1	$Q_{01}C_{01}-Q_{12}C_1+Q_{31}C_3=0$
2	$Q_{12}C_1-Q_{23}C_2+Q_{02}C_{02}=0$
3	$Q_{23}C_2-Q_{33}C_3-Q_{31}C_3=0$

In order to better understand how Eq. (5.5) can be used to solve linear algebraic equations, Example 5.1 presents a practical situation.

Example 5.1 Let us imagine three perfectly stirred tanks, interconnected as per Fig. 5.1. This system is also presented in Chapra and Canale (2005).

At the beginning, there is the same volume of pure water in all tanks, and pure water is fed into tanks 1 and 2 at rates Q_{01} and Q_{02} (m³/min), respectively. This system has a recycle, and part of the effluent of tank 3 returns to tank 1, at a rate Q_{31} (m³/min). During the entire process, the volume of liquid in the three tanks remains constant. The amounts of liquid that leave tanks 1 and 2 to enter tanks 2 and 3, respectively, are Q_{12} and Q_{23} (m³/min), and the amount of water that leaves tank 3 is Q_{33} (m³/min). Table 5.1 shows numerical values for all flow rates.

At a certain point, the streams Q_{01} and Q_{02} start feeding tanks 1 and 2 with a NaOH solution with concentrations of 10 mol/m³ (C_{01}) and 1 mol/m³ (C_{02}), respectively, instead of pure water, although all flow rates remain the same. What are the NaOH concentrations in all of the tanks when a steady state is reached?

Solution:

The application of conservation law for the three tanks in steady state will yield $E - L = 0$. The NaOH concentrations in tanks 1, 2, and 3 will be defined as C_1, C_2, and C_3, respectively. In this way, the mass balance for the three tanks can be written generating the linear algebraic equations system shown in Table 5.2.

Assuming the numerical values for all flow rates (Table 5.1) and for C_{01} and C_{02}, the equations in Table 5.2 become:

$$-7C_1 + 2C_3 = -50$$
$$7C_1 - 8C_2 = -1$$
$$8C_2 - 8C_3 = 0$$

The matrix form for this system of linear algebraic equations can be written as follows:

$[A]\{C\} = \{B\}$, in which:

$$[A] = \begin{bmatrix} -7 & 0 & 2 \\ 7 & -8 & 0 \\ 0 & 8 & -8 \end{bmatrix}, \{C\} = \begin{Bmatrix} C_1 \\ C_2 \\ C_3 \end{Bmatrix} \text{ and } \{B\} = \begin{Bmatrix} -50 \\ -1 \\ 0 \end{Bmatrix}$$

Keeping in mind the Eq. (5.5), the concentrations in the three tanks can be obtained by solving the expression below:

$$\{C\} = [A]^{-1}\{B\} \tag{5.6}$$

At this point, it is necessary to know how to handle inversion and multiplication of a matrix, and we will do that using Excel. In this book, we will use Excel 2016; however, there is not much difference among versions, and you can get some tips by using the F1 command (Help) in your Excel spreadsheet if you are using a different version of Excel.

In order to invert a matrix, it has to be written in the cells of the Excel spreadsheet, as shown on the left side of Fig. 5.2. After that, select the space in which you want to enter the inverted matrix (see the right side of Fig. 5.2), type in the function space at the top of your spreadsheet =MINVERSE(then select the matrix you want to invert (the one on the left side) and observe in the function space that the notation C4:E6 (the original matrix you want to invert) will appear.

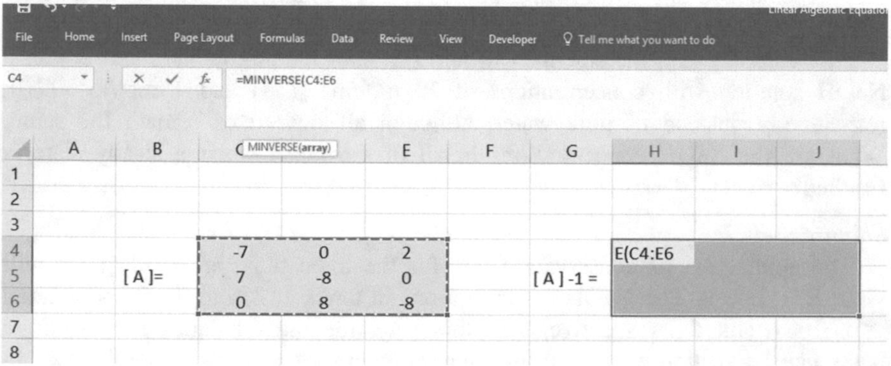

Fig. 5.2 Inverting a matrix using Excel

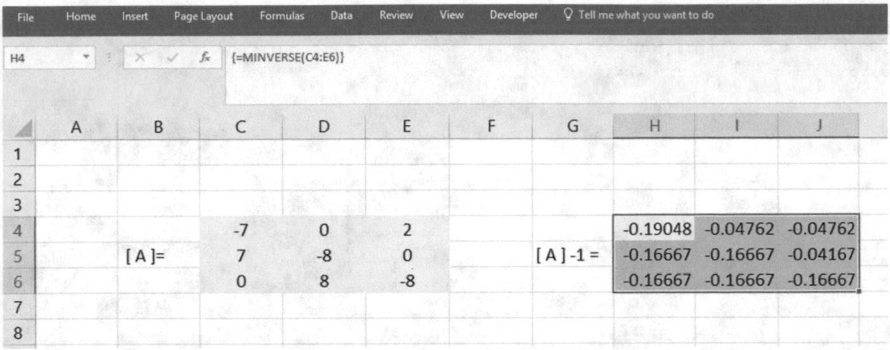

Fig. 5.3 Matrix A inverted, using Excel 2016

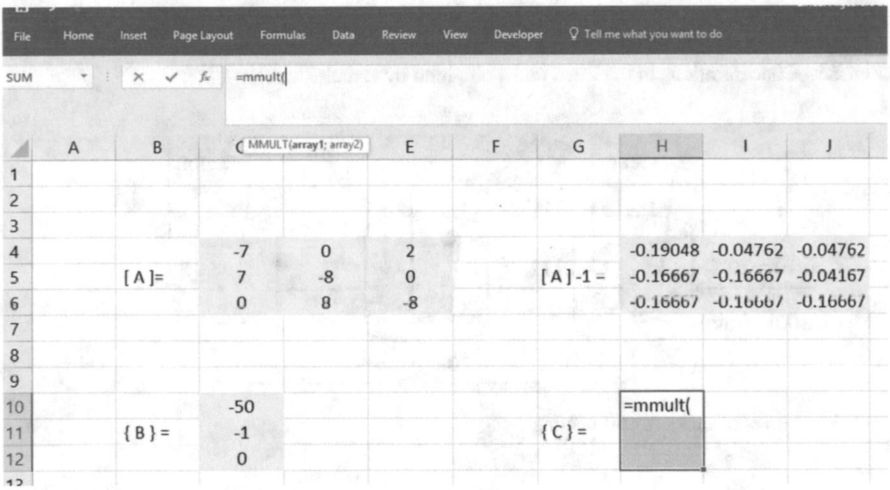

Fig. 5.4 Multiplying a matrix in Excel

Finally, close the parentheses, keep the *Ctrl* and *Shift* keys pressed, and press *Enter*. The inverted matrix will appear, as shown in Fig. 5.3.

In order to obtain $\{C\}$ (the concentrations in the three tanks), $[A]^{-1}$ (the inverted matrix of A) has to be multiplied by vector $\{B\}$, as per Eq. (5.6). To do that in Excel, follow these steps:

(1) Write vector $\{B\}$, as per the left side of Fig. 5.4.
(2) Select the cells where vector $\{C\}$ has to be written (see the right side of Fig. 5.4).
(3) In the function space (at the top of the spreadsheet in Fig. 5.4), type $=MMULT($ then select the matrix $[A]^{-1}$, followed by a semicolon, select the vector $\{B\}$, and close the parentheses, as per Fig. 5.5.
(4) Keep the *Ctrl* and *Shift* keys pressed, and press *Enter*. The vector C will appear, as shown in Fig. 5.5.

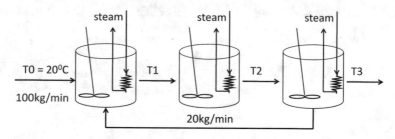

Fig. 5.5 Concentrations in the three tanks obtained by Excel

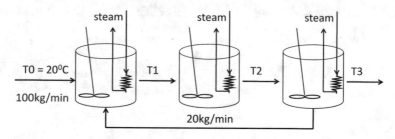

Fig. 5.6 Series of tanks for oil heating

Easily, we can obtain the concentrations in tanks 1, 2, and 3 in a steady state ($C_1 = 9.571$ and $C_2 = C_3 = 8.500$).

After learning how to solve linear algebraic equations using Excel, let us solve one more example, but this time considering the energy balance.

Example 5.2 Figure 5.6 shows three tanks in series used to preheat a multicomponent oil solution before it is fed into a distillation column for separation. This system was presented in Chap. 3 (Proposed Problem 3.7), but here we will consider the system in a steady state.

Saturated steam at a temperature of 250 °C condenses within a coil immersed in each tank. The oil is fed into the first tank at the rate of 100 kg min^{-1} and there is a recycle of 20 kg min^{-1} from tank 3 to tank 1. The flow rates into the second and the third tanks are 120 kg min^{-1}, since we assume that the volumes of the three tanks do not change over time. We also assume that the tanks are well mixed, so the temperature inside each tank is uniform, and that the specific heat, c_p, of the oil for the three tanks is 2.0 kJ kg^{-1} °C^{-1}.

Table 5.3 Energy balance for the three tanks

Tank	Enters (kJ/min)	Leaves (kJ/min)	Exchanges heat by convection (kJ/min)
1	$100\, c_p T_0 + 20\, c_p T_3$	$120\, c_p T_1$	$UA\,(T_{steam} - T_1)$
2	$120\, c_p T_1$	$120\, c_p T_2$	$UA\,(T_{steam} - T_2)$
3	$120\, c_p T_2$	$100\, c_p T_3 + 20\, c_p T_3$	$UA\,(T_{steam} - T_3)$

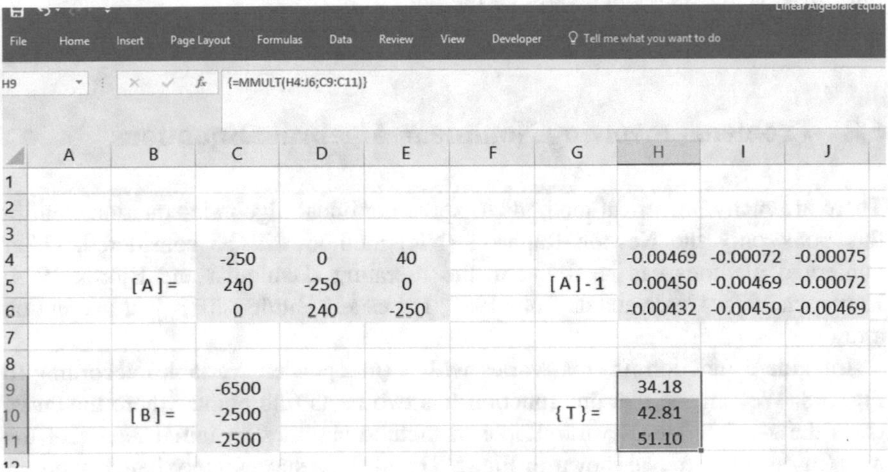

Fig. 5.7 Solution for the series of three tanks for oil heating using Excel

For each tank, there is heat exchange between oil and steam, and the product of the heat transfer coefficient (U) and the area (A) of the coil is 10 kJ min^{-1} °C^{-1} ($UA = 10$ kJ min^{-1} °C^{-1}).

The conservation law applied to each tank yields E − L = 0. Table 5.3 shows all terms needed to build the energy balance for the three tanks.

The linear algebraic equations that represent the energy balance for the series of three tanks can be seen below.

$$c_p(100T_0 + 20T_3 - 120T_1) + UA(T_{steam} - T_1) = 0$$
$$c_p(120T_1 - 120T_2) + UA(T_{steam} - T_2) = 0$$
$$c_p(120T_2 - 100T_3 - 20T_3) + UA(T_{steam} - T_3) = 0$$

Considering numerical values and rearranging terms, the final linear algebraic equations are:

$$-250T_1 + 40T_3 = -6500$$
$$240T_1 - 250T_2 = -2500$$
$$240T_2 - 250T_3 = -2500$$

Figure 5.7 shows the solution for this linear problem of energy balance using Excel. It can be observed that in a steady state, the temperatures in tanks 1, 2, and 3 are 34.18, 42.81, and 51.10 °C, respectively.

Observe that the energy balance for this system does not depend on the tanks' temperature at the beginning (at time equal to zero). That means that after a steady state is reached, the temperatures inside the tanks will always be the ones shown in Fig. 5.7, no matter what the initial conditions are. Moreover, it can be observed that the volume of the tanks is also not present in the energy balance. We can imagine that the smaller the volumes, the faster a steady state is reached, but the temperatures in a steady state will always be the ones in Fig. 5.7.

5.2 Problems Involving Nonlinear Algebraic Equations

There are many numerical methods to solve nonlinear algebraic equations, but in this book only the Newton–Raphson (NR) method will be considered. Other numerical methods can be found in the literature (Dahlquist and Björck 1974; Hornbeck 1975; Conte and de Boor 1980; Faires and Burden 2013, just to mention a few).

Imagine a function $f(x)$ that varies with x (independent variable) according to Fig. 5.8. We can see that this function has two roots (the points where the curve cross the x-axis). The Newton–Raphson method assumes an initial guess for the root (in our case, x_0, as shown in Fig. 5.8) and finds successively better approximations of the root (x_1, x_2, etc).

Starting with the initial guess x_0, the function is approximated by its tangent line, which intercepts the x-axis at x_1. The value x_1 is a better approximation of the root than the initial guess x_0. We can repeat this procedure one more time to obtain x_2, even closer to the root. The NR method is an iterative method, and after a few iterations, the root can be found.

Imagine an angle α generated from a tangent line drawn from the initial guess x_0. Numerically, the NR method can be easily represented applying the concept of the derivative and tangent.

Fig. 5.8 Visualization of the Newton–Raphson (NR) method to find the root of the function $f(x)$

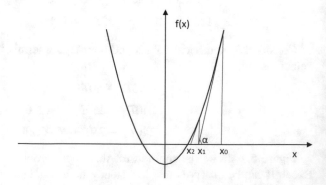

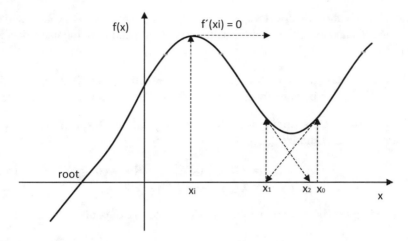

Fig. 5.9 Situations in which the Newton–Raphson (NR) method can fail. *Left*: $f'(x_i)$ is very close to zero. *Right*: The method oscillates around a local minimum

$$\tan \alpha = \frac{f(x_0) - 0}{x_0 - x_1} = f'(x_0) \quad \text{so} \quad x_1 = x_0 - \frac{f(x_0)}{f'(x_0)} \tag{5.7}$$

The NR method requires a good initial guess and finds just one root, usually one closer to the initial guess. Figure 5.9 shows some situations in which the NR method can fail. Observe that if in any iteration, $f'(x_i)$ is close to zero, the method does not converge (observe Fig. 5.9 and the denominator in Eq. 5.7). The method can also oscillate around a local minimum (see the right side of Fig. 5.9).

Section 5.2.1 will demonstrate how to apply the NR method to solve nonlinear algebraic equations systems.

5.2.1   Demonstration of the NR Method to Solve a Nonlinear Algebraic Equations System

Consider the continuous stirred tank reactor (CSTR) presented in Example 3.7. The system operates in a steady state regime and the reaction $A + B \xrightarrow{k} C$ takes place. In order to make the demonstration of the NR method more didactic, let us consider an isothermal CSTR, so the energy balance is not necessary. Assume that the CSTR operates at a constant temperature of 307 K. Moreover, let us consider mass balances only for reactants A and B, which are rewritten as follows:

Mass balance for A (mol/s): $Q(C_{A_{in}} - C_A) - kC_AC_BV = 0$ (3.12)

Mass balance for D (mol/s): $Q(C_{B_{in}} - C_B) - kC_AC_BV = 0$ (3.13)

Considering the numerical values for $C_{A_{in}}$, $C_{B_{in}}$, and Q (Table 3.2), and for V (Table 3.1), and assuming that k is equal to 0.06 m^3/mol min, the mass balances become:

$$\text{Mass balance for A (mol/s)} : C_A + 0.06\, C_A C_B = 200 \qquad (5.8)$$
$$\text{Mass balance for B (mol/s)} : C_B + 0.06\, C_A C_B = 200 \qquad (5.9)$$

We will start demonstrating the NR method by calling Eqs. (5.8) and (5.9) as functions $u(C_A, C_B)$ and $v(C_A, C_B)$, respectively.

$$u(C_A, C_B) = C_A + 0.06\, C_A C_B - 200 = 0 \qquad (5.10)$$
$$v(C_A, C_B) = C_B + 0.06\, C_A C_B - 200 = 0 \qquad (5.11)$$

Expanding functions u and v in the Taylor series, we obtain:

$$u_{i+1} = u_i + \left(C_{A_{i+1}} - C_{A_i}\right)\left(\frac{\partial u}{\partial C_A}\right)_i + \left(C_{B_{i+1}} - C_{B_i}\right)\left(\frac{\partial u}{\partial C_B}\right)_i \qquad (5.12)$$

$$v_{i+1} = v_i + \left(C_{A_{i+1}} - C_{A_i}\right)\left(\frac{\partial v}{\partial C_A}\right)_i + \left(C_{B_{i+1}} - C_{B_i}\right)\left(\frac{\partial v}{\partial C_B}\right)_i \qquad (5.13)$$

Remember that the estimates of the roots C_A and C_B correspond, according to the NR method, to the values of C_A and C_B that make u_{i+1} and v_{i+1} equal to zero. In this way, considering $u_{i+1} = v_{i+1} = 0$ and rearranging Eqs. (5.12) and (5.13), we obtain:

$$\frac{\partial u_i}{\partial C_A}C_{A_{i+1}} + \frac{\partial u_i}{\partial C_B}C_{B_{i+1}} = -u_i + C_{A_i}\frac{\partial u_i}{\partial C_A} + C_{B_i}\frac{\partial u_i}{\partial C_B} \qquad (5.14)$$

$$\frac{\partial v_i}{\partial C_A}C_{A_{i+1}} + \frac{\partial v_i}{\partial C_B}C_{B_{i+1}} = -v_i + C_{A_i}\frac{\partial v_i}{\partial C_A} + C_{B_i}\frac{\partial v_i}{\partial C_B} \qquad (5.15)$$

Equations (5.14) and (5.15) can be written in a matrix form, as follows:

$$\begin{bmatrix} \dfrac{\partial u_i}{\partial C_A} & \dfrac{\partial u_i}{\partial C_B} \\[2ex] \dfrac{\partial v_i}{\partial C_A} & \dfrac{\partial v_i}{\partial C_B} \end{bmatrix} \begin{Bmatrix} C_{A_{i+1}} \\ C_{B_{i+1}} \end{Bmatrix} = -\begin{Bmatrix} u_i \\ v_i \end{Bmatrix} + \begin{bmatrix} \dfrac{\partial u_i}{\partial C_A} & \dfrac{\partial u_i}{\partial C_B} \\[2ex] \dfrac{\partial v_i}{\partial C_A} & \dfrac{\partial v_i}{\partial C_B} \end{bmatrix} \begin{Bmatrix} C_{A_i} \\ C_{B_i} \end{Bmatrix}$$

or:

$$[Z]_i \{C\}_{i+1} = -\{F\}_i + [Z]_i \{C\}_i \qquad (5.16)$$

in which:

$$\begin{bmatrix} \dfrac{\partial u_i}{\partial C_A} & \dfrac{\partial u_i}{\partial C_B} \\[2ex] \dfrac{\partial v_i}{\partial C_A} & \dfrac{\partial v_i}{\partial C_B} \end{bmatrix} = [Z]_i = \text{Jacobian Matrix in } i$$

$$\left\{ \begin{matrix} u_i \\ v_i \end{matrix} \right\} = \{F\}_i = \text{Function at } i$$

$$\left\{ \begin{matrix} C_{A_i} \\ C_{B_i} \end{matrix} \right\} = \{C\}_i = \text{Initial guess in } i$$

$$\left\{ \begin{matrix} C_{A_{i+1}} \\ C_{B_{i+1}} \end{matrix} \right\} = \{C\}_{i+1} = \text{NR Prediction in } i+1$$

Observe that, if initial guesses for the concentrations of A and B ($\{C\}_i$) are given, the matrix $[Z]_i$ and the vector $\{F\}_i$ can be calculated. In this way, the concentrations of A and B for the next iteration ($\{C\}_{i+1}$) can be obtained by Eq. (5.16).

If the matrix $[Z]_i$ and the vector $\{C\}_i$ are multiplied and the result is added to $-\{F\}_i$, the right side of Eq. (5.16) can be written in a simpler way and Eq. (5.16) can be expressed by:

$$[Z]_i \{C\}_{i+1} = \{B\}_i \tag{5.17}$$

in which:

$$\{B\}_i = -\{F\}_i + [Z]_i \{C\}_i$$

Equation (5.17) represents a linear algebraic equations system and can be solved by multiplying both sides of Eq. (5.17) by the inverse matrix of $[Z]_i$, as presented in Sect. 5.1, to yield:

$$\{C\}_{i+1} = [Z]_i^{-1} \{B\}_i \tag{5.18}$$

The Newton–Raphson method transforms nonlinear to linear algebraic equations, and the solution for the linearized system can be calculated as presented in Sect. 5.1. Herein we have shown how the linearization is done; however, next time, to solve nonlinear algebraic equations, Eq. (5.16) can be used directly.

The Jacobian matrix $[Z]_i$ in Eq. (5.16) can be found by deriving Eqs. (5.10) and (5.11) with respect to C_A and C_B, to yield:

$$\frac{\partial u}{\partial C_A} = 1 + 0.06 C_B \tag{5.19}$$

$$\frac{\partial u}{\partial C_B} = 0.06 C_A \tag{5.20}$$

$$\frac{\partial v}{\partial C_A} = 0.06C_B \tag{5.21}$$

$$\frac{\partial v}{\partial C_B} = 1 + 0.06C_A \tag{5.22}$$

In order to solve Example 3.7 using the NR method, initial guesses are needed. In Table 3.2 it can be observed that A and B are fed into the reactor at a concentration of 200 mol/m^3. As they are reactants, they will be consumed, and the exit concentrations in a steady state must be lower than 200 mol/m^3. However, let us assume, as initial guesses (condition $i = 0$), an exit concentration of A and B equal to 200 mol/m^3. With this initial guess, all values with index i in Eq. (5.16) are known. $[Z]_0$ can be calculated by Eqs. (5.19), (5.20), (5.21), and (5.22), $\{F\}_0$ is obtained using Eqs. (5.10) and (5.11), and $\{C\}_0$ is our own initial guess. The only term unknown in Eq. (5.16) is $\{C\}_{i+1}$, which in our case is $\{C\}_1$, and represents the concentration of A and B for the next iteration. In this way, Eq. (5.16) can be written as follows:

$$\begin{bmatrix} 13 & 12 \\ 12 & 13 \end{bmatrix} \begin{Bmatrix} C_{A_1} \\ C_{B_1} \end{Bmatrix} = -\begin{Bmatrix} 2400 \\ 2400 \end{Bmatrix} + \begin{bmatrix} 13 & 12 \\ 12 & 13 \end{bmatrix} \begin{Bmatrix} 200 \\ 200 \end{Bmatrix}$$

After rearranging the right side of the equation, we obtain the linear algebraic equations system below, which can be solved using the numerical procedure presented in Sect. 5.1.

$$\begin{bmatrix} 13 & 12 \\ 12 & 13 \end{bmatrix} \begin{Bmatrix} C_{A_1} \\ C_{B_1} \end{Bmatrix} = \begin{Bmatrix} 2600 \\ 2600 \end{Bmatrix}$$

After the first calculation, the concentrations of C_A and C_B will be $C_{A_1} = C_{B_1} = 104$ mol/m3. After five iterations, the method converges and provides the concentrations of the reactants in a steady state. Table 5.4 shows the results for all iterations until convergence.

Sometimes it is not easy to analytically derive equations of the model to obtain the Jacobian matrix. When this is the case, numerical differentiation can be used, as presented in Sect. 5.2.2.

Table 5.4 Results from the Newton–Raphson (NR) method applied to Example 3.7

	Number of iterations						
	0	1	2	3	4	5	6
C_A (mol/m^3)	200	104	62.98	51.18	51.01	50.00	50.00
C_B (mol/m^3)	200	104	62.98	51.18	51.01	50.00	50.00

5.2.2 Numerical Differentiation

The numerical differentiation can be easily understood using the concept of a derivative. Consider a function $f(x)$ varying with x. The derivative of $f(x)$ with respect to x can be represented by:

$$\frac{df(x)}{dx} = f'(x) = \lim_{\Delta x \to 0} \frac{f(x + \Delta x) - f(x)}{\Delta x} \qquad (5.23)$$

Let us understand how to calculate the numerical derivative using a simple example: find the derivative of $f(x) = x^2$ at $x = 2$. The analytical derivative of this function can be easily obtained: $f'(x) = 2x$, and at $x = 2$, $f'(2) = 4$. In order to calculate the numerical derivative of this function, let us assume different values for Δx and apply Eq. (5.23).

$$\text{For } \Delta x = 0.1: \qquad f'(2) = \frac{(2 + 0.1)^2 - (2)^2}{0.1} = 4.1$$

$$\text{For } \Delta x = 0.01: \qquad f'(2) = \frac{(2 + 0.01)^2 - (2)^2}{0.01} = 4.01$$

$$\text{For } \Delta x = 0.001: \qquad f'(2) = \frac{(2 + 0.001)^2 - (2)^2}{0.001} = 4.001$$

We can observe that the smaller Δx is, the more precise the numerical derivative is.

The numerical derivative presented herein is the *forward difference* formula. It is also possible to obtain *backward* and *centered difference* formulas. Chapter 7 will present these two other possibilities, which can also be used to obtain the Jacobian matrix.

In the example of the CSTR presented previously, the Jacobian Z could be obtained numerically. Considering $\Delta x = 0.001$, and the functions in Eqs. (5.10) and (5.11), the Jacobian for the first iteration (initial guess $= C_A = C_B = 200 \, \text{mol/m}^3$) would be:

$$\frac{\partial u}{\partial C_A} = \frac{[(200 + 0.001) + 0.06(200 + 0.001)(200) - 200] - [200 + 0.06 \times 200 \times 200 - 200]}{0.001} = 13$$

$$(5.24)$$

$$\frac{\partial u}{\partial C_B} = \frac{[200 + 0.06(200)(200 + 0.001) - 200] - [200 + 0.06 \times 200 \times 200 - 200]}{0.001} = 12$$

$$(5.25)$$

$$\frac{\partial v}{\partial C_A} = \frac{[200 + 0.06(200 + 0.001)(200) - 200] - [200 + 0.06 \times 200 \times 200 - 200]}{0.001} = 12$$

$$(5.26)$$

$$\frac{\partial v}{\partial C_B} = \frac{[(200 + 0.001) + 0.06(200)(200 + 0.001) - 200] - [200 + 0.06 \times 200 \times 200 - 200]}{0.001} = 13$$

$$(5.27)$$

We can observe that the numerical derivative can be used successfully to obtain the Jacobian matrix, and this can be very helpful when an analytical derivative is difficult to obtain. Section 5.2.3 will apply Excel to solve nonlinear algebraic equations using the NR method.

5.2.3 Using Excel to Solve a Nonlinear Algebraic Equation Using the NR Method

In this section, we will use Excel to solve the nonlinear algebraic equations presented previously by Eqs. (5.10) and (5.11). Figure 5.10 presents a suggestion for a spreadsheet in Excel, in which cells in gray represent calculations done. In Fig. 5.10 the Jacobian matrix was calculated analytically.

The initial guesses and the value for the kinetic constant ($k = 0.06$) are written in line 8. Line 12 calculates Eqs. (5.10) and (5.11) using the first initial guess ($C_A = C_B = 200$). Lines 13 and 14 calculate the four partial derivatives to obtain the Jacobian matrix, also considering the initial guesses. To fill each of these six spaces in gray, click the mouse in the cell you want to calculate, then click the mouse in the *Insert Function* area and add an equals sign followed by the equations you want calculated (Fig. 5.10 shows an example for the cell that calculates $v(C_A, C_B)$). After that, press *Enter*.

In lines 17 and 18, $[Z]_i$, $[F]_i$ and $[C]_i$ are rewritten in a vector form, using the six values obtained previously and the initial guesses. To do that, click the mouse in the

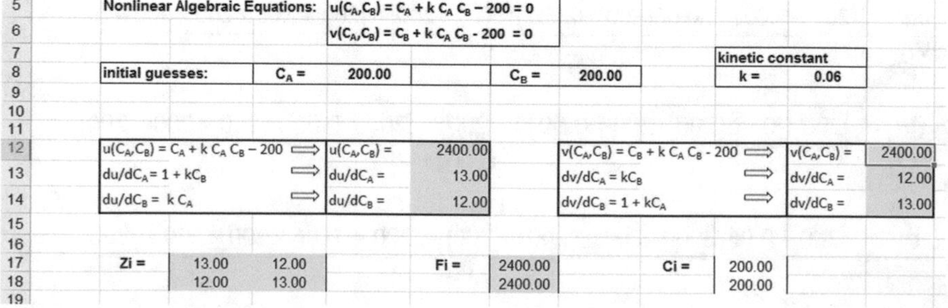

Fig. 5.10 Solving nonlinear algebraic equations using the Newton–Raphson (NR) method and Excel: calculating $[Z]_0$ and $\{F\}_0$

| File | Home | Insert | Page Layout | Formulas | Data | Review | View | Developer | ♀ Tell me what you want to do |

G21   ▾ : ✕ ✓ *fx* {=-G17:G18+MMULT(C17:D18;J17:J18)}

	A	B	C	D	E	F	G	H	I	J	K	L
1												
2		Solving nonlinear algebraic equations using NR method										
3												
4												
5		Nonlinear Algebraic Equations:			$u(C_A,C_B) = C_A + k\,C_A\,C_B - 200 = 0$							
6					$v(C_A,C_B) = C_B + k\,C_A\,C_B - 200 = 0$							
7									kinetic constant			
8		initial guesses:		$C_A =$	200.00		$C_B =$	200.00		$k =$	0.06	
9												
10												
11												
12		$u(C_A,C_B) = C_A + k\,C_A\,C_B - 200$ ⟹			$u(C_A,C_B) =$	2400.00		$v(C_A,C_B) = C_B + k\,C_A\,C_B - 200$ ⟹	$v(C_A,C_B) =$	2400.00		
13		$du/dC_A = 1 + kC_B$ ⟹			$du/dC_A =$	13.00		$dv/dC_A = kC_B$ ⟹	$dv/dC_A =$	12.00		
14		$du/dC_B = k\,C_A$ ⟹			$du/dC_B =$	12.00		$dv/dC_B = 1 + kC_A$ ⟹	$dv/dC_B =$	13.00		
15												
16												
17		Zi =	13.00	12.00		Fi =	2400.00		Ci =	200.00		
18			12.00	13.00			2400.00			200.00		
19												
20												
21		Vector {B} = - {F}i + [Z]i * {C}i			⟹	B =	2600.00					
22							2600.00					
23												
24		{C}ᵢ₊₁ = [Z]ᵢ⁻¹ {B}ᵢ			⟹	$[Z]_i^{-1} =$	0.52	-0.48				
25							-0.48	0.52				
26												
27			{C}i+1 =	104.00								
28				104.00								

Fig. 5.11 Solving nonlinear algebraic equations using the Newton–Raphson (NR) method and Excel: calculating $\{B\}_0$ and $\{C\}_1$

cell where you want to type, click again in the *Insert Function* area, type an equals sign, click in the corresponding cell, and press *Enter*.

Figure 5.11 shows how vector $\{B\}$ is obtained (see also Eq. 5.17). The space in which vector $\{B\}$ will be written must be selected using the mouse, then click the mouse in the *Insert Function* area, type an equals sign, type a minus sign, select all of the vector $[F]_i$ using the mouse, and add the multiplication of matrix $[Z]_i$ and $[C]_i$ (see details in Fig. 5.11).

The inverse of matrix $[Z]_i$ (represented by $[Z]_i^{-1}$) and the vector $\{C\}_{i+1}$ (the next guess) are obtained as was done in Sect. 5.1 (for linear algebraic equations problems). In this first iteration, the concentrations of A and B change from 200 to 104 mol/m^3.

Now we use the value 104 as our next guess, as shown in Fig. 5.12. In order to do that, we have to type the number *104* in the cells for the initial guesses (*E8* and *H8*), and all the calculations for this new guess will be done automatically because we write a genetic algorithm (see Fig. 5.12 with new values for the second guess). This next iteration brings the concentrations for A and B even closer to the real ones (in a steady state, $C_A = C_B = 50$ mol/m^3). In three more iterations, the roots for the nonlinear algebraic system are obtained. At the bottom of Fig. 5.12, a table has been built with the concentrations of A and B until the convergence is obtained.

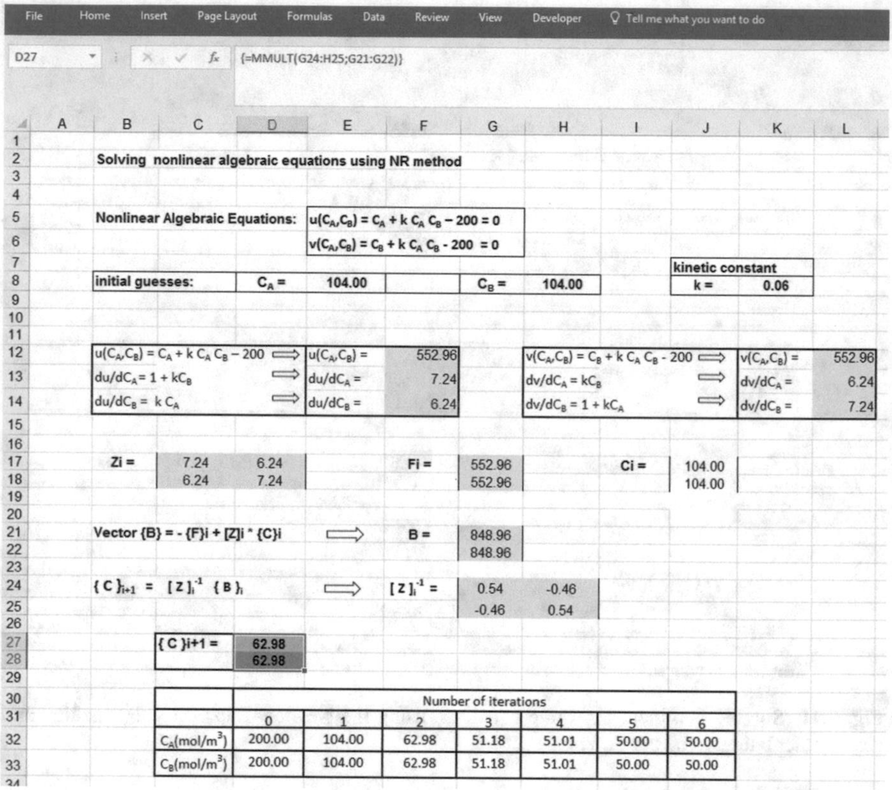

Fig. 5.12 Solving nonlinear algebraic equations using the Newton–Raphson (NR) method and Excel: changing initial guesses for the next iteration

Observe in Figs. 5.10, 5.11, and 5.12 that the Jacobian matrix (matrix $[Z]_i$) is calculated analytically. If numerical differentiation is needed, in the four cells (F13, F14, L13 and L14) related to $\partial u/\partial C_A$, $\partial u/\partial C_B$, $\partial v/\partial C_A$, and $\partial u/\partial C_B$, Eqs. (5.24), (5.25), (5.26), and (5.27) should be used instead of Eqs. (5.19), (5.20), (5.21), and (5.22).

5.3 Solving Linear and Nonlinear Algebraic Equations Using the *Solver* Tool

Try to again solve the problem of Sect. 5.2 imagining that the matrix $[Z]_i$ in Fig. 5.11 is:

$$\begin{bmatrix} 13 & 13 \\ 12 & 12 \end{bmatrix} = [Z]_i$$

You will see that it is impossible to invert the matrix $[Z]_i$, because its determinant is null. For a linear or nonlinear equations system, this situation can occur and the two methods presented in Sects. 5.1 and 5.2 cannot be used. The third approach presented in this section allows us to solve linear and nonlinear algebraic equations systems using the *Solver* tool in Excel.

The *Solver* command appears in Excel's *Data* tab. If it is not present in your computer, you have to activate it by following these steps:

(1) Click on the *File* tab, click on *Options*, and then click on the *Add-ins* category.
(2) In the *Manage* box, click on *Excel Add-ins*, and then click on *Go* (the *Add-ins* dialog box appears).
(3) In the *Add-ins available* box, select *Solver*, and click on *OK*.

The *Solver* tool finds an optimal value for one cell in a worksheet by changing the values of some cells that you specify. To understand better, let us imagine the nonlinear algebraic system solved in Sect. 5.2 and rewritten below:

$$u(C_A, C_B) = C_A + 0.06\, C_A C_B - 200 = 0 \qquad (5.10)$$

$$v(C_A, C_B) = C_B + 0.06\, C_A C_B - 200 = 0 \qquad (5.11)$$

The roots (C_A and C_B) for this system of equation will make $u(C_A, C_B) = v(C_A, C_B) = 0$. *Solver* can optimize just one cell, so we will optimize the cell that represents the summation of $u(C_A, C_B)$ and $v(C_A, C_B)$, by making it equal to zero. In fact, we usually use the sum of the *SQUARE* of the functions. This is because, in some situations, functions can assume positive or negative values, and their summation can be zero even if individually their values are far from zero. So, we will use $u^2 + v^2$ and not $u + v$ for the cell to be optimized.

In our case, the changing cells will be the cells related to C_A and C_B. Figure 5.13 shows the *Solver* tool being applied. The optimized and changing cells are in gray. The initial guesses for C_A and C_B are the same as those considered previously ($C_A = C_B = 200$ mol/m^3).

Observe in the function space (at the top of the spreadsheet in Fig. 5.13) how u^2 was calculated. The cells *E7* and *E8* are used to represent C_A and C_B, respectively. Cell *C12* represents the cell to be optimized. When clicking on *Data* and then *Solver*, the screen at the right side of Fig. 5.13 appears. In the *Set Objective* option, click on the cell *C12*. Then choose the *Value Of* option and type *0* [i.e., a zero] in the space straight ahead. By doing it this way, we force the functions $u(C_A, C_B)$ and $v(C_A, C_B)$ to go to zero, and it is possible because *Solver* will change cells *E7* and *E8* until it finds values for C_A and C_B (the roots) that make $u(C_A, C_B)$ and $v(C_A, C_B)$ equal to zero. In this way, in the *By Changing Variable Cells* option, click on the cells *E7* and *E8*. After that, click on *Solver* and the results shown in Fig. 5.14 will appear. Finally, you can click on *OK*. Observe that the optimized cell (*C12*) is very close to zero, and $C_A = C_B = 50.0$ mol/m^3, as expected (see Fig. 5.14).

Observe in Fig. 5.13 that there are three options for the optimization method. We use the generalized reduction gradient (GRG) nonlinear method because our system

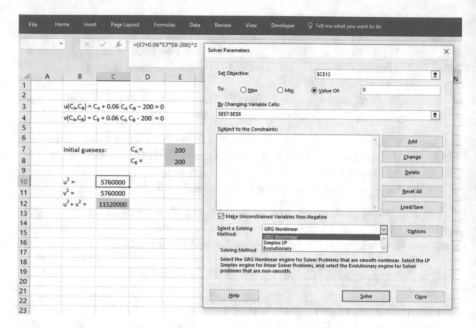

Fig. 5.13 Use of the *Solver* tool to solve nonlinear algebraic Eqs. (5.10) and (5.11)

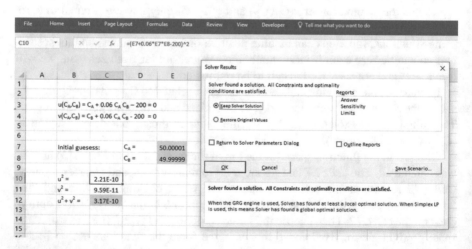

Fig. 5.14 Solution of nonlinear algebraic Eqs. (5.10) and (5.11) obtained by the *Solver* tool

is nonlinear and smooth. It is not within the scope of this book to deal with optimization methods. The reader can find books specific to the optimization issue in the literature (Foulds 1981; Nocedal and Wright 1999; Fletcher 2000; Rao 2009; Edgar and Himmelblau 2001, etc.).

Our example could be solved in a different way, by using the *Subject to the Constraints* box (see this box in Fig. 5.13). We could optimize one of the functions

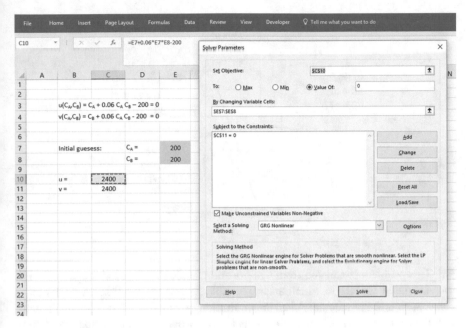

Fig. 5.15 Alternative way to solve Eqs. (5.10) and (5.11) using the *Solver* tool

(say, $u(C_A, C_B)$) and restrict the other cell ($v(C_A, C_B)$) to zero, using the *Subject to the Constraints* box. Figure 5.15 shows this other possibility.

Usually the approach using the *Solver* tool in Excel is much simpler than the NR method; however, sometimes it does not work well, especially when the equation system is very nonlinear. This is the case with the Arrhenius expression to represent the kinetic constants ($k = k_0 \exp\left(-\frac{E_A}{RT}\right)$). In a nonisothermal reactor, a not very good initial guess for temperature can make your problem reach wrong values for the roots (local minima). This will be observed in Proposed Problem 5.4.

The three alternatives presented in this chapter to solve algebraic equations systems allow us to deal with a great number of problems. The next two chapters are dedicated to ordinary and partial differential equations (ODEs and PDEs), respectively, so more complex problems will be able to be solved.

Proposed Problems

5.1) Imagine a tank in which three streams coming from different parts of a chemical plant are mixed. The mass fractions of compounds A, B, and C in each stream, as well as their mass flow rates, are shown in the figure below. Find the mass flow rate for each stream (F_1, F_2, and F_3) that should be used to obtain an exit stream with the composition shown in the figure (30% of A, 40% of B, and 30% of C). Write the mass balance for each compound (A, B, and C) and solve the linear algebraic equations system using the procedures shown in Sects. 5.1 and 5.3. Compare the results. Assume that the system is in a steady state.

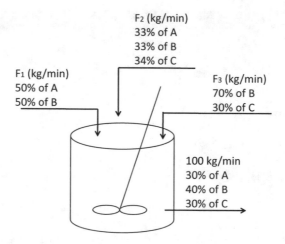

5.2) This problem is adapted from Chapra and Canale (2005), studying a system of four insulated CSTRs in which an irreversible first-order reaction (A → B) takes place. A solution of reactant A in a concentration of 1 mol L^{-1} feeds the first reactor at a flow rate of 10 L/h. There are recycles from reactor 3 to reactor 2 and from reactor 4 to reactor 3, with flow rates of 5 L/h and 3 L/h, respectively.

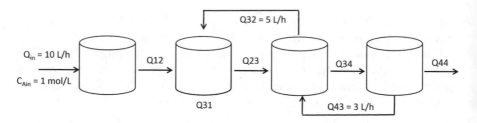

The reactors have different volumes and operate at different temperatures. The table below shows the volume and the kinetic constant for each reactor.

Reactor	V (L)	k (1/h)
1	25	0.075
2	75	0.15
3	100	0.4
4	25	0.1

Use the methods seen in Sects. 5.1 and 5.3 to find the concentration of reactant A in each reactor in a steady state. Compare the results obtained from the two different methods. After solving your problem, make some changes in your system (one at a time) and observe the results (suggestion: increase the values for k or reduce the concentration of A fed into reactor 1).

5.3) Consider Proposed Problem 3.8 in which the irreversible reaction $A + B \xrightarrow{k} C$ $+D$ takes place in an isothermal CSTR, but this time assume that the reactor operates in a steady state. The rate constant k is equal to 0.855 l/mol s. A solution with reactants A and B was added to the reactor at a flow rate (F) of 5 L min^{-1} and at concentrations of A and B equal to 0.7 and 0.4 mol L^{-1}, respectively $(C_{A_1} = 0.7$ mol L^{-1} and $C_{B_1} = 0.4$ mol L^{-1}). There are no products C and D being fed into the reactor $(C_{C_1} = C_{D_1} = 0)$. The outlet volumetric flow rate is also 5 L min^{-1}, and the volume of liquid inside the reactor remains equal to 40 L over the entire reaction. The model that represents this system can be seen below:

$$F(C_{A_1} - C_A) - 0.855C_A C_B V = 0$$
$$F(C_{B_1} - C_B) - 0.855C_A C_B V = 0$$
$$F(C_{C_1} - C_C) + 0.855C_A C_B V = 0$$
$$F(C_{D_1} - C_D) + 0.855C_A C_B V = 0$$

Solve the equations system that represents this CSTR and find the concentrations of A, B, C, and D in a steady state. Solve this problem considering the methods shown in Sects. 5.2 and 5.3, and compare the results.

5.4) Consider Example 3.7 for a CSTR with a cooling jacket operating in a steady state in which the exothermic reaction $A + B \xrightarrow{k} C$ takes place. The mass and energy balances that represent this reactor are rewritten as follows:

Mass balance for A (mol/min): $Q(C_{Ain} - C_A) - kC_A C_B V = 0$ (3.12)

Mass balance for B (mol/min): $Q(C_{Bin} - C_B) - kC_A C_B V = 0$ (3.13)

Mass balance for C (mol/min): $Q(C_{Cin} - C_C) + kC_A C_B V = 0$ (3.14)

Energy balance for the reactor (J/min): $Q\rho c_p(T_{in} - T) + UA(Tj - T)$
$$+ kC_A C_B V(-\Delta H)_R = 0$$ (3.15)

Energy for the cooling fluid (J/min): $Q_j \rho_j c_{p_j}(Tj_{in} - Tj) + UA(T - Tj) = 0$ (3.16)

Using the numerical values in Tables 3.1 and 3.2, solve the nonlinear algebraic equations system using the NR approach (Sect. 5.2) and obtain the concentrations of A, B, and C, and the temperatures of the reactor and the cooling jacket in a steady state. Try to solve this problem using the *Solver* tool (Sect. 5.3) and observe that this approach is not robust (due to a highly nonlinear characteristic of this system).

5.5) Imagine a liquid–liquid extractor of five stages, as depicted by the figure below. A liquid solution containing some component to be recovered (solute) is fed into the left side of the extractor. This solution contains a mass fraction of the solute equal to Y_{in} and flows at F_1 (kg/h). A solute-free liquid (solvent) is fed into the right side of the extractor (at countercurrent flow) at a mass flow rate equal to F_2 (kg/h). This solvent is immiscible with the liquid fed into the left side and will recover the solute due to the relative solubility of the solute in these two immiscible liquids. Assume that the extractor operates in a steady state and that there is a liquid–liquid equilibrium at each stage i given by:

$$K = X_i/Y_i$$

in which:

$K =$ distribution coefficient
$X_i =$ mass fraction of the solute in the solvent
$Y_i =$ mass fraction of the solute in the liquid to be recovered
$i =$ number of the stage

Assume the following numerical values:

F_1 (kg/h)	F_2 (kg/h)	Y_{in}	X_{in}	K
500	1000	0.3	0.0	4

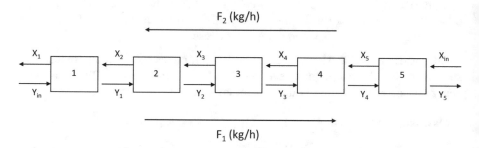

(a) Calculate the mass balance for each stage and find the following system of algebraic equations to represent the extractor:

$$
\begin{aligned}
i = 1: &\quad Y_{in} - A\,Y_1 + B\,Y_2 = 0 \\
i = 2: &\quad Y_1 - A\,Y_2 + B\,Y_3 = 0 \\
i = 3: &\quad Y_2 - A\,Y_3 + B\,Y_4 = 0 \\
i = 4: &\quad Y_3 - A\,Y_4 + B\,Y_5 = 0 \\
i = 5: &\quad Y_4 - A\,Y_5 + B\,X_{in} = 0
\end{aligned}
$$

in which:

$$A = \left(\frac{F_2}{F_1}K\right) + 1$$

$$B = \frac{F_2}{F_1}K$$

(b) Solve the algebraic equations system obtained in item (a) using the approaches presented in Sects. 5.1 and 5.3, and find the mass fraction of the solute in both streams leaving the extractor. Do you think that the separation was effective?

References

Burden, R., Faires, J.D., Burden, A.M.: Numerical Analysis. Cengage Learning, Boston (2014)

Chapra, C.C., Canale, R.P.: Numerical Methods for Engineers, 5th edn. Mc Graw Hill, New York (2005)

Conte, S.D., de Boor, C.: Elementary Numerical Analysis: An Algorithm Approach, 3rd edn. McGraw-Hill Book Company, New York (1980)

Dahlquist, G., Björck, A.: Numerical Methods. Prentice-Hall, Inc., Englewood Cliffs (1974)

Edgar, T.F., Himmelblau, D.M.: Optimization of Chemical Processes, 2nd edn. McGraw Hill Chemical Engineering Series, New York (2001)

Faires, J.D., Burden, R.L.: Numerical Methods, 4th edn. Brooks/Cole Cengage Learning, Boston (2013)

Fletcher, R.: Practical Methods of Optimization, 2nd edn. Wiley, New York (2000)

Foulds, L.R.: Optimization Techniques: An Introduction. Springer, New York (1981)

Hornbeck, R.W.: Numerical Methods. Quantum Publishers, Inc., New York (1975)

Isaacson, E., Keller, H.B.: Analysis of Numerical Method. Wiley, New York (1966)

Nocedal, J., Wright, S.J.: Numerical Optimization. Springer-Verlag, New York (1999)

Rao, S.S.: Engineering Optimization: Theory and Practice, 4th edn. Wiley, Hoboken (2009)

Chapter 6
Solving an Ordinary Differential Equations System

In ordinary differential equations (ODEs), dependent variables (such as temperature, concentration, etc.) vary with only one independent variable (a spatial variable or time). In this way, all lumped-parameter problems in a transient regime, as well as all distributed-parameter problems in a steady state varying with just one of the three spatial variables, are described by ODEs.

Numerical methods for integration of ODEs can be classified as explicit and implicit. In explicit methods, the state of a system in a condition $(i + 1)$ is calculated based only on the previous condition (i) (a sequential solution). In implicit methods, the solution in $i + 1$ depends on the state of both i and $i + 1$ (a simultaneous solution), being more difficult to implement. Implicit methods are very useful for stiff systems (in which a dependent variable varies abruptly with an independent variable) because they are numerically more stable. Most systems in chemical plants are not stiff if operated in planned conditions; therefore, only explicit methods are presented this book. The reader can see details about implicit numerical methods elsewhere (Davis 1984; Rao 2002, etc.).

6.1 Motivation

Imagine the derivative shown in Eq. (6.1) with the correspondent boundary condition:

$$\frac{dy}{dx} = \cos(x) + 1 \tag{6.1}$$

Boundary condition: at $x = 0$, $y(0) = 0$

The original version of this chapter was revised. An erratum to this chapter can be found at
https://doi.org/10.1007/978-3-319-66047-9_8

113
L.M.F. Lona, *A Step by Step Approach to the Modeling of Chemical Engineering Processes*, https://doi.org/10.1007/978-3-319-66047-9_6

In this case, if one wants to know, for example, the value of y when x is equal to 10, Eq. (6.1) can be analytically integrated to generate:

$$y(x) = \sin(x) + x \qquad (6.2)$$

which gives us, at $x = 10$, $y(10) = 9.456$.

In this case, Eq. (6.1) is very easy to integrate; however, when systems with many complex ODEs must be solved simultaneously, numerical solution can be an interesting alternative. Before presenting the numerical methods, let us take a look at the expression that represents the expansion of the Taylor series (Eq. 6.3):

$$y_{i+1} = y_i + h\left(\frac{dy}{dx}\right)_i$$

$$+ \frac{1}{2!}h^2\left(\frac{d^2y}{dx^2}\right)_i + \frac{1}{3!}h^3\left(\frac{d^3y}{dx^3}\right)_i + \ldots + \frac{1}{n!}h^n\left(\frac{d^ny}{dx^n}\right)_i + Rn \qquad (6.3)$$

in which:

$y =$ function that depends on x

$x =$ independent variable

$h = x_{i+1} - x_i = \Delta x =$ infinitesimal variation in the independent variable

$n =$ number of terms in the Taylor series

$Rn =$ error after n terms

The solution to Eq. (6.1) could also be obtained using expansion of the Taylor series. The more terms that are considered, the more precise the solution is. Let us solve Eq. (6.1) considering n (number of terms in the Taylor series) as equal to 1, 2, 3, and 4, and compare the results. We can assume $h = x_{i+1} - x_i = \Delta x = 1$. Since the condition at $y(0)$ is known, we can calculate $y(1)$, $y(2)$, and so on until we discover the value of $y(10)$. The disadvantage in using expansion of the Taylor series is the calculation of higher-order derivatives when using more than one term ($n > 1$), which, in some cases, can be a hard task. In this example, the higher-order derivatives for Eq. (6.1) are easily obtained, as follows:

$$\frac{dy}{dx} = \cos(x) + 1 \qquad (6.1)$$

$$\frac{d^2y}{dx^2} = -\sin(x) \qquad (6.4)$$

$$\frac{d^3y}{dx^3} = -\cos(x) \qquad (6.5)$$

$$\frac{d^4y}{dx^4} = \sin(x) \qquad (6.6)$$

So the truncated Taylor series for $n = 1$, 2, 3, and 4, neglecting Rn, can be represented by Eqs. 6.7, 6.8, 6.9 and 6.10 respectively:

$$y_{i+1} = y_i + h[\cos(x_i) + 1] \qquad (6.7)$$

$$y_{i+1} = y_i + h[\cos(x_i) + 1] + \frac{1}{2!}h^2[-\sin(x_i)] \qquad (6.8)$$

$$y_{i+1} = y_i + h[\cos(x_i) + 1] + \frac{1}{2!}h^2[-\sin(x_i)] + \frac{1}{3!}h^3[-\cos(x_i)] \quad (6.9)$$

$$y_{i+1} - y_i + h[\cos(x_i) + 1]$$

$$+ \frac{1}{2!}h^2[-\sin(x_i)] + \frac{1}{3!}h^3[-\cos(x_i)] + \frac{1}{4!}h^4[\sin(x_i)] \quad (6.10)$$

Starting from the condition at $x = 0$ ($y(0) = 0$) and considering $h = \Delta x = 1$, it is possible to calculate $y(1)$. From the $y(1)$, we calculate $y(2)$ and so on until we reach $y(10)$. Figure 6.1 compares the analytical solution of Eq. (6.1) with the results obtained from the Taylor series using a different number of terms.

We can indeed verify from Fig. 6.1 that the increase in the number of terms generates more accurate results; however, it is interesting to point out that if a lower increment h were considered, the curves obtained from the Taylor series would be closer to the analytical solution.

If, on one hand, more precise results are obtained for bigger n values, on the other hand the more terms in the series result in more difficult calculation, because additional derivatives of higher order are needed. In this way, sometimes the Taylor series does not represent a real advantage compared with the analytical solution which demands calculation of just one integral to generate the exact solution.

Most researchers consider Taylor series methods too expensive for most practical problems. In this way, there is another class of numerical methods that imitate the Taylor series methods, without the necessity for calculating higher-order derivatives. These are called Runge–Kutta methods and will be presented in Sect. 2.

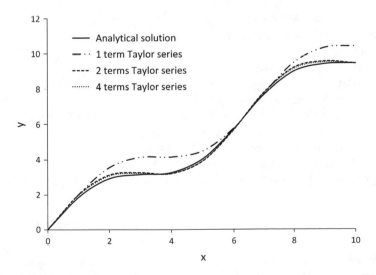

Fig. 6.1 Comparison between an analytical solution and solutions obtained by an expanded Taylor series truncated after 1, 2, and 4 terms

6.2   Runge–Kutta Numerical Methods

The general formula for the Runge–Kutta (RK) family of methods is:

$$y_{i+1} = y_i + \sum_{j=1}^{v} \omega_j K_j \qquad (6.11)$$

In which:

$$K_j = hf\left(x_i + c_j h, \; y_i + \sum_{l=1}^{j-1} a_{jl} K_l\right) \qquad (6.12)$$

$$c_1 = 0$$

In the RK formulas, the term f represents the derivative of the dependent variable:

$$f(x_i, y_i) = \left(\frac{dy}{dx}\right)_i \qquad (6.13)$$

and:

y_i = dependent variable in condition i

y_{i+1} = dependent variable in condition $i + 1$

j and l = counters in the summation

ω, a, and c = parameters of RK methods

h = increment

v = order of RK methods

The order of the Runge–Kutta method will depend on the number of terms considered in the summation of Eq. (6.11), as follows:

$v = 1$: First order Runge–Kutta method (RK1), also called the *Euler method*

$v = 2$: Second order Runge–Kutta method (RK2)

$v = 3$: Third order Runge–Kutta method (RK3)

$v = 4$: Forth order Runge–Kutta method (RK4)

Or generalizing: If $v = j$, a jth order Runge–Kutta method is generated.

6.2.1   First Order Runge–Kutta Method, or Euler Method

To find out the formula for the first order *Runge–Kutta* (*Euler* method), $v = 1$ is considered in Eq. (6.11). If we assume that $\omega_1 = 1$, the following formula for the Euler method is obtained:

$$\boxed{\begin{aligned} y_{i+1} &= y_i + K_1 \\ K_1 &= hf(x_i, y_i) \end{aligned}}$$

$$\Longrightarrow \quad \text{RK1 – Euler}$$

(6.14)

(6.15)

or substituting Eq. (6.15) in Eq. (6.14):

$$y_{i+1} = y_i + hf(x_i, y_i) \tag{6.16}$$

Equation (6.16) is exactly the expansion of the Taylor series presented in Eq. (6.3) considering $n = 1$. The Euler method can be easily visualized in Fig. 2.2 in Chap. 2, redrawn below (see Fig. 6.2).

If a condition (x_0, y_0) is given and if Δx is sufficiently small $(y_1 \cong v_1)$, the value of the function y at $x_1 = x_0 + \Delta x$ can be obtained considering the definitions of the tangent and derivative.

$$\tan \alpha = \frac{y_1 - y_0}{x_1 - x_0} = \left(\frac{dy}{dx}\right)_0 \tag{6.17}$$

So :

$$y_1 = y_0 + \Delta x \left(\frac{dy}{dx}\right)_0 \tag{6.18}$$

or generalizing :

$$y_{i+1} = y_i + \Delta x \left(\frac{dy}{dx}\right)_i \tag{6.19}$$

6.2.2 Second Order Runge–Kutta Method

To obtain the formula for the second order Runge–Kutta method (RK2), $v = 2$ is adopted in Eq. (6.11) to yield:

$$y_{i+1} = y_i + \omega_1 K_1 + \omega_2 K_2 \tag{6.20}$$

$$K_1 = hf(x_i, y_i) \tag{6.21}$$

$$K_2 = hf(x_i + c_2 h, y_i + a_{21} K_1) \tag{6.22}$$

The terms ω_1, ω_2, c_2, and a_{21} in Eqs. (6.20) and (6.22) must be found. If we assume that $c_2 h$ and $a_{21} K_1$ in Eq. (6.22) are sufficiently small, $f(x_i + c_2 h, y_i + a_{21} K_1)$ can be represented by an expansion of the Taylor series (we will adopt $n = 1$), and Eq. (6.22) can be written as:

$$K_2 = h\left[f(x_i, y_i) + c_2 hf'_x(x_i, y_i) + a_{21} K_1 f'_y(x_i, y_i)\right] \tag{6.23}$$

in which:

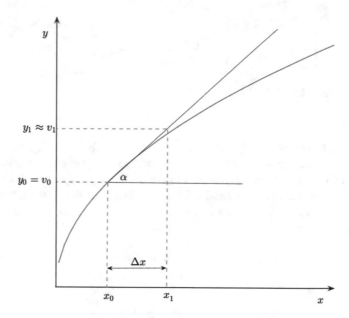

Fig. 6.2 Visualization of the Euler (first-order Runge–Kutta (RK1)) method

$$f(x_i, y_i) = \left(\frac{dy}{dx}\right)_i \qquad (6.13)$$

$$f'_x(x_i, y_i) = \left(\frac{\partial f}{\partial x}\right)_i \qquad (6.24)$$

$$f'_y(x_i, y_i) = \left(\frac{\partial f}{\partial y}\right)_i \qquad (6.25)$$

Substituting Eqs. (6.21) and (6.23) in (6.20) and rearranging, we obtain:

$$\boxed{y_{i+1} = y_i + (\omega_1 + \omega_2)hf(x_i,y_i) + \omega_2 h^2\left[c_2 f'_x(x_i,y_i) + a_{21}f'_y(x_i,y_i)f(x_i,y_i)\right]} \quad (6.26)$$

On the other hand, if we represent y_{i+1} using expansion of the Taylor series with $n = 2$, we get (compare this with Eq. 6.3):

$$y_{i+1} = y_i + h\left(\frac{dy}{dx}\right)_i + \frac{1}{2!}h^2\left(\frac{d^2 y}{dx^2}\right)_i + R_2 \qquad (6.27)$$

Taking into account Eq. (6.13) and considering the R_2 error as being very small, Eq. (6.27) can be written as:

$$y_{i+1} = y_i + hf(x_i, y_i) + \frac{1}{2!}h^2 f'(x_i, y_i) \tag{6.28}$$

By definition:

$$f'(x_i, y_i) = \left(\frac{df}{dx}\right)_i = \left(\frac{\partial f}{\partial x}\right)_i + \left(\frac{\partial f}{\partial y}\right)_i\left(\frac{dy}{dx}\right)_i \tag{6.29}$$

Substituting Eq. (6.29) in Eq. (6.28) yields:

$$y_{i+1} = y_i + hf(x_i, y_i) + \frac{1}{2!}h^2\left[\left(\frac{\partial f}{\partial x}\right)_i + \left(\frac{\partial f}{\partial y}\right)_i\left(\frac{dy}{dx}\right)_i\right] \tag{6.30}$$

Using Eqs. (6.13), (6.24), and (6.25), Eq. (6.30) can also be written as:

$$\boxed{y_{i+1} = y_i + hf(x_i, y_i) + \frac{1}{2}h^2\left[f'_x(x_i, y_i) + f'_y(x_i, y_i)f(x_i, y_i)\right]} \tag{6.31}$$

Observe the similarity between Eqs. (6.26) and (6.31). The terms ω_1, ω_2, c_2, and a_{21} in RK2 can be found by comparing Eq. (6.26) (generated from the application of RK2) with Eq. (6.31) (obtained from the application of the Taylor series truncated in the second term, $n = 2$). We conclude that:

$$\omega_1 + \omega_2 = 1 \tag{6.32}$$

$$\omega_2 c_2 = 1/2 \tag{6.33}$$

$$\omega_2 a_{21} = 1/2 \tag{6.34}$$

Since there are four unknown variables (ω_1, ω_2, c_2, and a_{21}) and only three equations (Eqs. 6.32, 6.33, 6.34), one of the variables is fixed to allow calculation of the others. Two examples are shown:

Case 1 Fix $c_2 = 0.5$ so $\omega_2 = 1$, $\omega_1 = 0$, and $a_{21} = 0.5$, and Eqs. (6.20), (6.21), (6.22), which represent RK2, can be expressed as follows:

$$y_{i+1} = y_i + K_2 \tag{6.35}$$

$$K_1 = hf(x_i, y_i) \tag{6.15}$$

$$K_2 = hf(x_i + 0.5h, y_i + 0.5K_1) \tag{6.38}$$

RK2

Case 2 Fix $c_2 = 1.0$ so $\omega_2 = 0.5$, $\omega_1 = 0.5$, and $a_{21} = 1.0$, and the equations for RK2 will be:

$$y_{i+1} = y_i + 0.5(K_1 + K_2) \qquad\qquad\qquad\qquad (6.37)$$

$$K_1 = hf(x_i, y_i) \qquad\qquad\text{RK2}\qquad\qquad (6.15)$$

$$K_2 = hf(x_i + h, y_i + K_1) \qquad\qquad\qquad\qquad (6.38)$$

For cases 1 and 2 of RK2, only first-order derivatives are needed. This is an advantage compared with the Taylor series truncated in the second term, which demands also the second-order derivative. In cases 1 and 2, K_1 represents the first-order derivative applied in the condition (x_i, y_i), as occurred in the Euler method. In cases 1 and 2, K_2 represents the first-order derivatives in the condition $(x_i + 0.5h, y_i + 0.5K_1)$ and $(x_i + h, y_i + K_1)$, respectively, so K_2 can only be obtained after K_1 is calculated.

In a similar manner, Runge–Kutta methods of higher order can be developed. Observe that the RK methods are based on the Taylor series, so *the higher the Runge–Kutta order is, the more accurate the method is, because more terms in the Taylor series are considered.* From the practical point of view, the most commonly used type of Runge–Kutta method is the fourth order Runge–Kutta (RK4), presented in Sect. 6.2.3 (RK methods of other orders can be seen in in may books, such as Davis 1984; Varma and Morbidelli 1997; Rao 2002; and Chapra and Canale 2005).

6.2.3 Runge–Kutta Method of the Fourth Order

The expression for the fourth-order Runge–Kutta method (RK4) can be obtained as was done for the case of RK2, and can be seen as follows:

$$y_{i+1} = y_i + \frac{h}{6}(K_1 + 2K_2 + 2K_3 + K_4) \qquad\qquad (6.39)$$

$$K_1 = f(x_i, y_i) \qquad\qquad\qquad\qquad (6.40)$$

$$K_2 = f(x_i + 0.5h, \ y_i + 0.5hK_1) \qquad\qquad\text{RK4}\qquad\qquad (6.41)$$

$$K_3 = f(x_i + 0.5h, \ y_i + 0.5hK_2) \qquad\qquad\qquad\qquad (6.42)$$

$$K_4 = f(x_i + h, \ y_i + hK_3) \qquad\qquad\qquad\qquad (6.43)$$

RK4 demands the calculation of four first derivatives $(K_1, K_2, K_3,$ and $K_4)$ at four different points for independent and dependent variables (x, y). Figure 6.3 compares the results obtained from the solution of Eq. (6.1) using RK methods of different orders with the analytical solution.

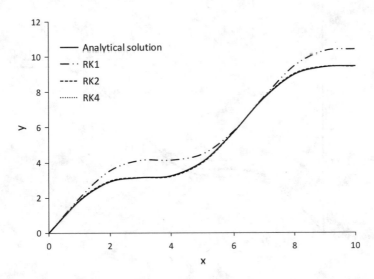

Fig. 6.3 Comparison of an analytical solution and solutions obtained by Runge–Kutta methods of the first, second, and fourth order, considering $h = 1$

As mentioned earlier, the higher the order of RK method is, the more precise the results are. This is more evident for nonsmooth functions. Observe in Fig. 6.3 that the RK1 method is closer to the analytical solution in the regions in which y varies in a more linear way with x (from 0 to 1 and from 6 to 7). If the function is not very nonlinear, the Euler method can be used successfully if a low value of h (step) is adopted.

The choice of the ideal step (h) is of crucial importance to obtain reliable results, but how do we pick the best step if we do not know the analytical solution to compare? The idea is to use different steps and compare the results. Fig. 6.4 solves Eq. (6.1) using the Euler method for different values of h. Observe that the curves are practically coincident for $h = 0.05$ and $h = 0.1$, so there is no advantage in using steps lower than 0.05.

Observe that to obtain almost the same accuracy in the results, RK4 used a step more than ten times bigger than the one used in the Euler method (compare Figs. 6.3 and 6.4).

Section 6.3 and 6.4 will show how to solve both a single ODE or a system of ODEs using an Excel spreadsheet and Visual Basic for Applications (VBA), respectively. The procedure adopted in Sect. 6.3 can be tedious and laborious; however, it can be very useful in order to clearly understand how the calculus is done in RK methods, especially for systems of equations being simultaneously solved.

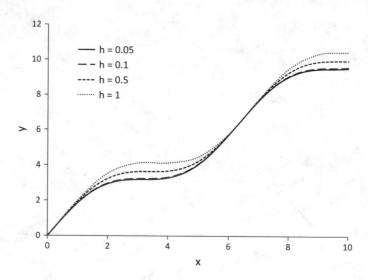

Fig. 6.4 Comparison between different values of steps for the solution of Eq. (6.1) using the Euler method

6.3 Solving ODEs Using an Excel Spreadsheet

Section 6.3.1 will use an Excel spreadsheet to solve just one ODE, and Sect. 6.3.2 will solve a system with more than one interdependent ODE.

6.3.1 Solving a Single ODE Using Runge–Kutta Methods

In order to understand how to solve an ODE using a spreadsheet in Excel, let us start with the very simple example presented in Example 4.1. The equation below represents the axial increase in temperature, in a steady state, of a liquid flowing in a cylindrical tube of 60 m, exchanging heat with a jacket.

$$\frac{dT}{dx} = \frac{U2\pi R}{Q\rho c_p}(300 - \mathrm{T}), \quad \text{with the boundary condition: at } x = 0, \ T = 20°C$$

This equation can be rewritten as:

$$f(x, T) = \frac{dT}{dx} = M(300 - \mathrm{T}), \text{ at } x_0 = 0, T_0 = 20°C \qquad (6.44)$$

in which:

$M = U2\pi R/(Q\rho c_p) = $ constant

$R = $ radius $ = 0.2$ m

U = global coefficient of heat transfer = 60,000 J/h m^2 °C
Q = volumetric flow rate = 4 m^3/h
ρ = density = 900 kg/m^3
c_p = specific heat of the fluid = 3000 J/kg °C

From a didactic point of view, let us start solving this ODE using the Euler method.

6.3.1.1 Euler Method

A suggestion for a spreadsheet using Euler method to solve Eq. (6.44) can be seen in Fig. 6.5, where in the first lines the ODE and the boundary condition are printed just as a comment. Lines 10 to 12 show the parameter of the model, needed to calculate the constant M. To calculate M, select a cell (in our case, the cell $C14$), go to the function space (at the top of your spreadsheet), type an equals sign followed by the expression that represents the constant M, and press *Enter* (in gray is written, as a comment, the expression you have to type). Then click on a cell in which to insert the step (in our case, the cell $C15$ was used and $h = 1$ was chosen). After that, write down the condition at x_0 (see cells $C18$ and $D18$ for x_0 and T_0, respectively).

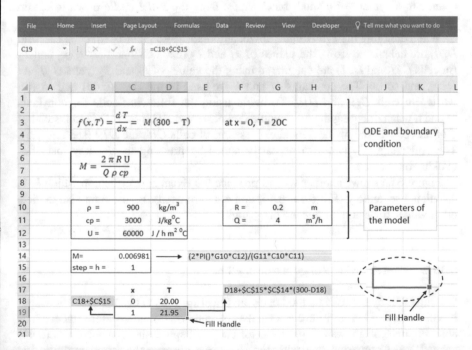

Fig. 6.5 Suggestion for an Excel spreadsheet to calculate a single ordinary differential equation (ODE) using the Euler method

According to the Euler method, the independent and dependent variables at x_1 can be calculated based on the condition at x_0 using Eqs. 6.45 to 6.48 (compare with Eqs. 6.14, 6.15, 6.16).

$$x_1 = x_0 + h \tag{6.45}$$
$$T_1 = T_0 + K_1 \tag{6.46}$$
$$K_1 = hf(x_0, T_0) \tag{6.47}$$

or:

$$T_1 = T_0 + hf(x_0, T_0) \tag{6.48}$$

To calculate x_1, select cell *C19*, type in the function space an equals signal followed by the corresponding equations (Eq. 6.45), and press *Enter*. In the same way, to calculate T_1, select cell *D19*, type in the function space an equal signals followed by the corresponding equations (Eq. 6.48), and press *Enter*. In Fig. 6.5, these two expressions are also written in gray as a comment.

After that, select together the two cells (*C19* and *D19*) containing the calculus you have just done for x_1 and T_1, and hover the cursor over the small box at the bottom right corner of the selected cells. This small box is called the *Fill Handle* (Fig. 6.5 shows a zoom of a cell inside the dashed line where the *Fill Handle* is indicated). When the mouse cursor is directly above the *Fill Handle*, the cursor will change to a symbol of a small black cross. Drag the *Fill Handle* down to obtain values of x and T for the next steps.

When dragging the *Fill Handle* down, automatically x_2 and T_2 (in cells *C20* and *D20*) are calculated using the values of x_1 and T_1 (cells *C19* and *D19*), x_3 and T_3 (in cells *C21* and *D21*) are calculated using the values of x_2 and T_2 (cells *C20* and *D20*), and so on. The constants M and h are used in the Eqs. (6.45) and (6.48), written in cells *C19* and *D19*. To avoid using the following cells for constants M and h when dragging the Fill Handle down, we use *C14* and *C15* instead of *C14* and *C15* in the mathematical expressions of cells *C19* and *D19* (see Fig. 6.5). The dollar sign (*$*) can also be easily added by pressing the *F4* key just after clicking on the cell.

Figure 6.6a shows another way to produce the Excel spreadshee, in which k_1 is calculated separately (Eq. 6.45, 6.46, and 6.47 are used). Figure 6.6a also depicts other values for x and T after the dragging.

6.3.1.2 Fourth order Runge Kutta method

Solving ODE by the RK4 method using an Excel spreadsheet is more laborious, but it will be shown herein in order to make the numerical method more understandable. Figure 6.6b shows a suggestion for an Excel spreadsheet to solve Eq. (6.44) using RK4. Comparing this with Fig. 6.6a, we observe that another three first derivatives are needed (K_2, K_3, and K_4).

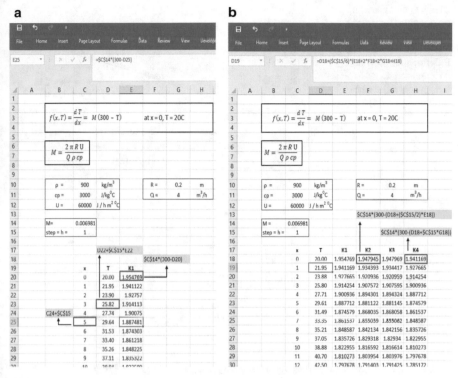

Fig. 6.6 Suggestion for an Excel spreadsheet to solve Eq. (6.44) using (**a**) the Euler method (*left*) and (**b**) the fourth-order Runge–Kutta (RK4) method (*right*)

Knowing the condition at (x_0, T_0), the condition at the first step (x_1, T_1) can be calculated according to the RK4 method as follows:

$$T_1 = T_0 + \frac{h}{6} (K_1 + 2K_2 + 2K_3 + K_4) \tag{6.49}$$
$$K_1 = f(x_0, T_0) \tag{6.50}$$
$$K_2 = f(x_0 + 0.5h, T_0 + 0.5hK_1) \tag{6.51}$$
$$K_3 = f(x_0 + 0.5h, T_0 + 0.5hK_2) \tag{6.52}$$
$$K_4 = f(x_0 + h, T_0 + hK_3) \tag{6.53}$$
$$x_1 = x_0 + h \tag{6.45}$$

The first-order derivatives K_1, K_2, K_3, and K_4 are calculated in cells $E18, F18$, $G18$ and $H18$, respectively. Figure 6.6b shows a detail for the expression used to calculate K_2 and K_4. In cell $C19$, x_1 is calculated as was done previously; in cell $D19$, T_1 is calculated using Eq. (6.49) (see detail in the function space in Fig. 6.6b). Then select (individually or in groups) the cells $E18, F18, G18, H18, C19$, and $D19$, and drag the *Fill Handle* down to obtain all values shown in Fig. 6.6b.

The temperature profiles obtained by the integration of Eq. (6.44) reveal an almost linear dependence between temperature and length for this specific system (Fig. 6.7), so, in this case, the difference between the Euler and RK4 methods is not very expressive, even for high values of integration steps (h).

Fig. 6.7 Comparison between the fourth-order Runge–Kutta (RK4) method (*down*) and the Euler method (*up*) for different values of integration steps when solving Eq. (6.44)

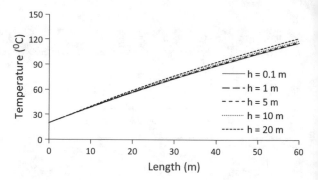

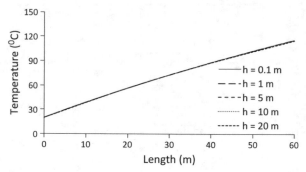

6.3.2   Solving a System of Interdependent ODEs Using Runge–Kutta Methods

In order to better understand how to apply RK methods to solve a system of ODEs, let us revisit Example 4.3, which modeled bitubular heat transfer operating in a concurrent way. The equations that represent this system are rewritten below:

Benzene : $W_{ben}cp_{ben}\dfrac{dT_{ben}}{dx} = U\ 1.25\ \pi(T_{tol} - T_{ben})$,   at $x = 0, T_{ben} = 60\,^\circ\text{F}$

Toluene : $W_{tol}\ cp_{tol}\dfrac{dT_{tol}}{dx} = U\ 1.25\ \pi\ (T_{ben} - T_{tol})$,   at $x = 0, T_{tol} = 170\,^\circ\text{F}$

This system can be also written as follows:

Benzene : $f(x, Tb, Tt) = \dfrac{dTb}{dx} = Mb(Tt - Tb)$,   at $x = 0, Tb = 60\,^\circ\text{F}$ (6.54)

Toluene : $g(x, Tb, Tt) = \dfrac{dTt}{dx} = Mt(Tb - Tt)$,   at $x = 0, Tt = 170\,^\circ\text{F}$ (6.55)

in which:

$Mb = 1.25\pi U/(W_{ben}cp_{ben}) = \text{constant}$

$Mt = 1.25\pi U/(W_{tol}cp_{tol}) = \text{constant}$

$W_{ben} = \text{mass flow of benzene} = 9820\ \text{lb/h}$

$W_{tol} = \text{mass flow of toluene} = 6330\ \text{lb/h}$

$cp_{ben} = \text{specific heat of benzene} = 0.425\ \text{Btu/(lb\ }^\circ\text{F)}$

cp_{tol} = specific heat of toluene = 0.440 Btu/(lb °F)
U = global coefficient of heat transfer = 0.8 Btu/(h in² °F)
$T_b - T_{ben}$ — benzene temperature flowing in the tube (°F)
$Tt = T_{tol}$ = toluene temperature flowing in the annulus (°F)
x = independent variable = heat transfer length (in)
Let us simultaneously solve Eq. (6.54 and 6.55) using first the Euler method.

6.3.2.1   Euler method

Knowing the condition at $x = 0$ (x_0, Tb_0 and Tt_0), the condition at the first step (x_1, Tb_1 and Tt_1) can be calculated as follows:

$$x_1 = x_0 + h \tag{6.45}$$
$$Tb_1 = Tb_0 + hK_1 \tag{6.56}$$
$$Tt_1 = Tt_0 + hL_1 \tag{6.57}$$
$$K_1 = f(x_0, Tb_0, Tt_0) \tag{6.58}$$
$$L_1 = g(x_0, Tb_0, Tt_0) \tag{6.59}$$

Observe that there is a first derivative for the dependent variable Tb (which we call K_1) and another for Tt (here named L_1). Note that K_1 and L_1 are interdependent (see Eqs. 6.54 and 6.55) and must be simultaneously solved. Fig. 6.8 shows a suggestion for an Excel spreadsheet to solve this system with two ODEs (Eqs. 6.54 and 6.55) using Euler method.

The constants Mb and Mt, in cells C15 and C16, are calculated based on the parameters written in lines 11 to 13. Figure 6.8 highlights the first calculations made for K_1, L_1, Tb, and Tt. After the first calculations are done, the *Fill Handle* must be dragged down to obtain how Tb and Tt vary with position, which, if plotted, generates the curves shown in Fig. 4.5.

6.3.2.2   RK4 method

If the RK4 method is used, the condition at the first step (x_1, Tb_1 and Tt_1) is obtained based on x_0, Tb_0 and Tt_0 (the boundary condition), using Eqs. (6.60 and 6.61):

$$Tb_1 = Tb_0 + \frac{h}{6} (K_1 + 2K_2 + 2K_3 + K_4) \tag{6.60}$$

$$Tt_1 = Tt_0 + \frac{h}{6} (L_1 + 2L_2 + 2L_3 + L_4) \tag{6.61}$$

in which the first derivatives for the functions Tb and Tt at $x = x_0$ are:

$$K_1 = f(x_0, Tb_0, Tt_0) \tag{6.58}$$
$$K_2 = f(x_0 + 0.5h, Tb_0 + 0.5hK_1, Tt_0 + 0.5hL_1) \tag{6.62}$$
$$K_3 = f(x_0 + 0.5h, Tb_0 + 0.5hK_2, Tt_0 + 0.5hL_2) \tag{6.63}$$

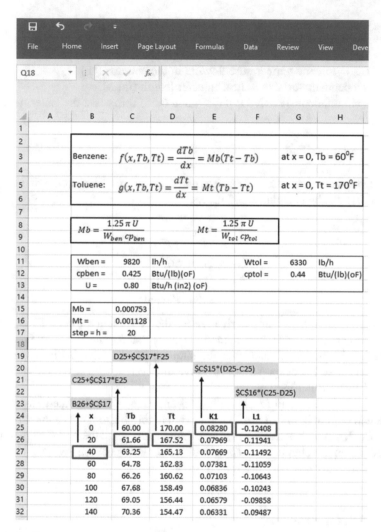

Fig. 6.8 Suggestion for an Excel spreadsheet to solve a system with two ordinary differential equations (ODEs) using the Euler method

$$K_4 = f(x_0 + h, Tb_0 + hK_3, Tt_0 + hL_3) \tag{6.64}$$
$$L_1 = g(x_0, Tb_0, Tt_0) \tag{6.59}$$
$$L_2 = g(x_0 + 0.5h, Tb_0 + 0.5hK_1, Tt_0 + 0.5hL_1) \tag{6.65}$$
$$L_3 = g(x_0 + 0.5h, Tb_0 + 0.5hK_2, Tt_0 + 0.5hL_2) \tag{6.66}$$
$$L_4 = g(x_0 + h, Tb_0 + hK_3, Tt_0 + hL_3) \tag{6.67}$$

Figure 6.9 presents a suggestion for how to solve simultaneously Eqs. 6.54 and 6.55 using the RK4 method in an Excel spreadsheet. The first calculations for Tb and Tt, as well as for some first derivatives, are highlighted.

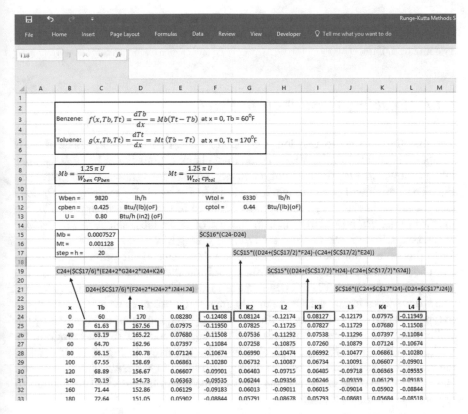

Fig. 6.9 Suggestion for an Excel spreadsheet to solve Eqs. (6.54 and 6.55) using the fourth-order Runge–Kutta (RK4) method

An Excel spreadsheet can be very useful to solve systems with very few equations by the Euler method, or to solve just one equation by RK4. For bigger systems, an Excel spreadsheet is impracticable, but it has been presented herein mainly for didactic reasons. The next section will show the procedure usually adopted to solve ODE systems.

6.4   Solving ODEs Using Visual Basic

The most effective way to solve a system of ODEs is to develop a computer code, which can be written in many different computer languages (Fortran, C, C++, Pascal, Visual Basic, Java Matlab, etc.). The focus of this book is to use Excel, therefore VBA (Visual Basic for Applications) will be the language presented in this section.

The idea of this book is not to show the VBA language in detail, because this can be seen in many specific books (Billo 2007; Walkenbach 2013a, b) and on internet sites. Herein will be presented examples of a few computer programs to solve ODEs using Euler and RK4 methods. These codes will be generic enough to allow students to use them as a starting point to develop their own programs for different systems of ODEs.

6.4.1 Enabling Visual Basic in Excel

The first thing to do to develop a computer code in Excel is to check if your computer is enabled to use VBA. To do that, check if your Excel has a *Developer* tab, as shown in Fig. 6.10. If not, click on the *File* tab, shown by the arrow in Fig. 6.10, which will open the window shown in Fig. 6.11. By clicking on *Options* (see the arrow in Fig. 6.11), the window depicted in Fig. 6.12 will appear. Choose the *Customize Ribbon* button and, in the list of main tabs, select the *Developer* check box. Click on *OK* to close the *Options* dialog box.

At this point your Excel is enabled to use the Visual Basic program. Now you can choose the *Developer* tab, and then click on the *Visual Basic* icon (see the dashed arrow in Fig. 6.10) to open a space in which you can write your program. The space shown in Fig. 6.13 is originally gray, but it becomes white (and available to type the code) after clicking on the *View Code* icon, indicated in Fig. 6.13 by the solid arrow. After typing your code into this space, you can run the program by clicking on the green triangle indicated by a dashed arrow in Fig. 6.13. If needed, the reader can see detail on how to debug a program in the *Help* option in Excel. However, if the code is correct, the click on the green triangle will be enough to provide the simulation results.

6.4.2 Developing an Algorithm to Solve One ODE Using the Euler Method

Like all programming languages, VBA provides many different possibilities of commands that can be used to develop a code. We will use just a few herein. Reader

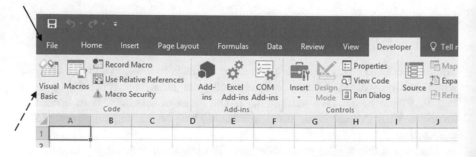

Fig. 6.10 Detail of the *Developer* tab and the *Visual Basic* icon in the Excel spreadsheet

Fig. 6.11 Detail of how to
open the Excel *Options*
window

can learn more about the different commands in VBA in the specific literature
(Billo 2007; Walkenbach 2013a, b, etc.).

To understand how to develop a program code, let us start with a very simple
problem, which can be represented by just one ODE, and let us adopt the Euler
method as the numerical method to solve this ODE. Proposed Problem 3.1 studies
the concentration of HCl inside a tank over time, for which modeling generates the
following ODE:

$$\frac{dC_{HCl}}{dt} = 0.01 - 0.4\, C_{HCl} \tag{6.68}$$

with the initial condition: at $t = 0$, $C_{HCl} = 0.01$ kg/m^3.

A suggestion of a simple code to solve Eq. (6.68) using the Euler method is
depicted in Fig. 6.14a. This code is written in the space that is used to write VBA
code in Excel, shown in Fig. 6.13. The name of the program is *Euler* and the word
Sub is used to start the code. When the expression *Sub Euler ()* is written and the

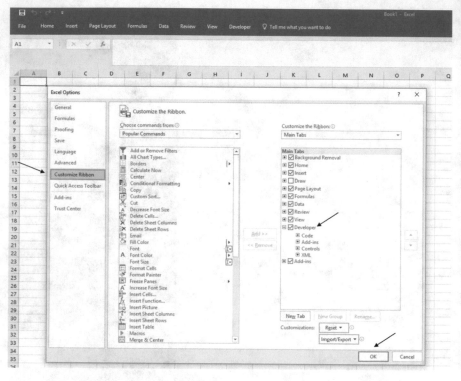

Fig. 6.12 Detail of the *Customize Ribbon* option, to activate the *Develop* tab in Excel

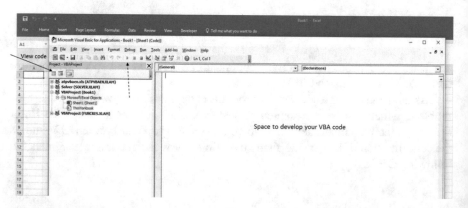

Fig. 6.13 Detail of the space in which to write Visual Basic (VBA) code in Excel

Enter key is pressed, automatically the expression *End Sub* will appear (see the last line of the code) and the words *Sub* and *End Sub* will turn blue. The code must be written between these two commands.

In the first three lines of the code, the initial condition (at $ti = 0$, $CHCl = 0.01$) and the final time (tf, at which we want to stop the simulation) are attributed. We

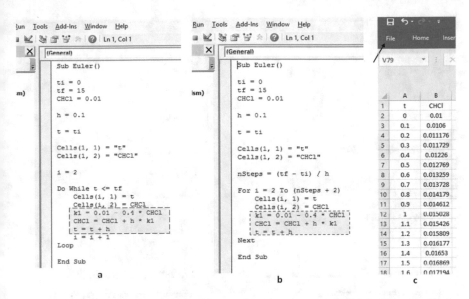

Fig. 6.14 **a, b** Suggestion for a code to solve Eq. (6.68) using **a** the *Do While* command and **b** the *For ... Next* command. **c** Results obtained using Visual Basic (VBA) code

choose to integrate Eq. (6.68) up to 15 hours, and we assume a step size (h) equal to 0.1 (see line 4 of the code).

VBA writes the results in the Excel spreadsheet, so to define the cells that will receive the results, we use the command *Cells (k,j)* = in which k indicates the line and j indicates the column of the spreadsheet. If we want to write a word, we must use quotation marks, as indicated in lines 6 and 7 of the code in Fig. 6.14a, which generates cells *A1* and *B1* in Fig. 6.14c.

The Euler method demands a loop, and in Fig. 6.14a this was done by the command *Do While*, which must be closed by the command *Loop* (see both in blue). Between these two commands the Euler method calculation is done, step by step, until it reaches *tf* equal to 15 hours. The calculation needed for the Euler method is highlighted in Fig. 6.14a (compare this with Eqs. (6.14) and (6.15)). Observe that we use the command *Cells (k,j)* inside the loop, so the values of t and *CHCl* for each loop will be written in the Excel spreadsheet (see the table generated in Fig. 6.14c). There is a counter (i) used to print the results, which started from 2 (see line 8 of the code), because line 1 of the spreadsheet is dedicated to the title of the table of results.

Another option for a VBA code to solve Eq. (6.68) using the Euler method is presented in Fig. 6.14b. Observe that the code is basically the same, but the command to build the loop is *For ... Next*. The variable *nSteps* is the number of integration steps. The calculus inside the loop is done 150 times (($tf - ti$)/h or (15-0)/0.1), but it uses *nSteps + 2* in the *For ... Next* command, because the counter i starts from 2.

These two codes could be written in a more general way if we create a *function*. Figure 6.15a shows an alternative for the code in Fig. 6.14a.

```
Sub Euler()                             Sub Euler()

ti = 0                                  xi = 0
tf = 15                                 xf = 15
CHCl = 0.01                             y = 0.01

h = 0.1                                 dx = 0.1

t = ti                                  x = xi

Cells(1, 1) = "t"                       Cells(1, 1) = "x"
Cells(1, 2) = "CHCl"                    Cells(1, 2) = "Y"

i = 2                    ⟹             i = 2

Do While t <= tf                        Do While x <= xf
    Cells(i, 1) = t                         Cells(i, 1) = x
    Cells(i, 2) = CHCl                       Cells(i, 2) = y
    Call RungeKutta1(t, CHCl, h)            Call RungeKutta1(x, y, dx)
    i = i + 1                               i = i + 1
Loop                                    Loop

End Sub                                 End Sub

Function Derivative(CHCl, k1)           Function Derivative(x, y, dydx)
k1 = 0.01 - 0.4 * CHCl                  dydx = 0.01 - 0.4 * y
End Function                            End Function

Function RungeKutta1(t, CHCl, h)        Function RungeKutta1(x, y, dx)
    Call Derivative(CHCl, k1)               Call Derivative(x, y, dydx)
    CHCl = CHCl + h * k1                     k1 = dydx
    t = t + h                               y = y + dx * k1
End Function                                x = x + dx
                                        End Function
```

a   b

Fig. 6.15 Use of the command *Function* when solving an ordinary differential equation (ODE) by the Euler method

Observe that the main program in Fig. 6.15a is very similar to the one shown in Fig. 6.14a, except that the calculus demanded by the Euler method is done in the function *RungeKutta1* (compare the highlighted commands in Figs. 6.14a and 6.15a).

The function *RungeKutta1* needs the step (h) and the previous values of t and *CHCl* to provide the next values of t and *CHCl*, which are sent to the main program. Since *CHCl*, t, and h are common between the main program and the function *RungeKutta1*, these three variables are passed and received as an *argument* (see the variable names, separated by commas, between the parentheses where *RungeKutta1* is called and in the function *RungeKutta1*).

The function *RungeKutta1* could calculate $k1$ inside it; however, we decide to call another function to calculate the derivative (see the function *Derivative*), which needs the *CHCl* value to return $k1$, so these two variables are also arguments (see *CHCl*, $k1$) in the function *Derivative* and where this function is called).

Figure 6.15b repeats the code presented in Fig. 6.15a but uses more generic nomenclature. From now on, the independent and dependent variables will be called x and y, respectively. Also, $k1$ and h will be replaced by $dydx$ and dx, respectively. Besides, the independent variable x will be passed as an argument in the function *Derivative* (see the arrows in Fig. 6.15b), because differently from Eq. (6.68), ODEs can also depend on an independent variable.

The code in Fig. 6.15b could be used to solve all systems represented by one ODE, just by changing the ODE in the function *Derivative* and the initial conditions in the main program.

Do not forget to save your program often. This is done by clicking on *File* as shown by the arrow in Fig. 6.14c. However, more importantly, do not forget to save your workbook as a *macro-enabled workbook* in VBA. Choose the option *Excel Macro-Enabled Workbook (*.xlsm)* when saving.

6.4.3 Developing an Algorithm to Solve One ODE Using the Runge–Kutta Fourth-Order Method

To solve Eq. (6.68) by the fourth-order Runge–Kutta (RK4) method, we can use the generic code shown in Fig. 6.15b but we exchange the function *Rungekutta1* for the function *RungeKutta4*, presented in Fig. 6.16. The RK4 method calculates the derivative four times $(K_1, K_2, K_3,$ and $K_4)$, at different values of x and y (independent and independent variables) as presented in Eqs. (6.40), (6.41), (6.42), (6.43). Because of this, the function *RungeKutta4* calls the function *Derivative* four times.

In Fig. 6.16, $k1$ is calculated as was done in Figure 6.15b; however, $k2$ must be calculated at $(x_i + 0.5h, y_i + 0.5hK_1)$ (see Eq. 6.41), so, just after the calculus of $k1$, we define the points x, y in which $k2$ must be calculated. These two values x and y are transitory (just to calculate $k2$), so they are called *xtran* and *ytran*, respectively. To calculate $k2$, the function *RungeKutta4* passes as an argument *xtran* and *ytran* when calling the function *Derivative* (see the command *Call Derivative (xtran, ytran, dydx)* in Fig. 6.16); however, the function *Derivative* can be written exactly as presented in Figure 6.15b, using *Function Derivative (x, y, dydx)*.

After obtaining $k2$, we need to find out *xtran* and *ytran* to calculate $k3$. We can see by Eq. (6.42) that *xtran* will be the same as was used to obtain $k2$, so it is not updated, but *ytran* will be $y_i + 0.5hk_2$ (see Eq. 6.42). The same procedure is followed until we obtain $k4$. After that, dependent and independent variables are updated (see the last lines of the code in Fig. 6.16 and Eq. (6.39) to obtain y_{i+1}).

```
Function RungeKutta4(x, y, dx)
' Calculate K1
    Call Derivative(x, y, dydx)
    k1 = dydx

' Define x and y to calculate K2
    ytran = y + k1 * 0.5 * dx
    xtran = x + 0.5 * dx

'Calculate K2
    Call Derivative(xtran, ytran, dydx)
    k2 = dydx

'Define x and y to calculate K3
    ytran = y + k2 * 0.5 * dx

'Calculate K3
    Call Derivative(xtran, ytran, dydx)
    k3 = dydx

'Define x and y to calculate K4
    ytran = y + k3 * dx
    xtran = x + dx

'Calculate K4
    Call Derivative(xtran, ytran, dydx)
    k4 = dydx

'Update dependent variable
y = y + (1 / 6) * (k1 + 2 * k2 + 2 * k3 + k4) * dx

'Update independent variable
    x = x + dx

End Function
```

Fig. 6.16 Example of a function for the fourth-order Runge–Kutta (RK4) method

6.4.4 Developing an Algorithm to Solve a System of ODEs Using the Euler and Fourth-Order Runge–Kutta Methods

In this section, let us see how to develop a code in VBA to solve a system of ODEs. We will use Example 3.8, which models a continuous stirred tank reactor (CSTR) with a cooling jacket operating in a transient regime. Equations (3.17, 3.18, 3.19, 3.20 and 3.21), representing the mass and energy balance of the reactor and the energy balance for the jacket, are rewritten as follows:

Mass balance for reactant A (mol):

$$\frac{dC_A}{dt} = \frac{Q}{V}(C_{Ain} - C_A) - \left[k_0 \exp\left(-\frac{E_A}{RT}\right)\right]C_A C_B \tag{6.69}$$

Mass balance for reactant B (mol):

$$\frac{dC_B}{dt} = \frac{Q}{V}(C_{Bin} - C_B) - \left[k_0 \exp\left(-\frac{E_A}{RT}\right)\right]C_A C_B \qquad (6.70)$$

Mass balance for product C (mol):

$$\frac{dC_C}{dt} = \frac{Q}{V}(C_{Cin} - C_C) + \left[k_0 \exp\left(-\frac{E_A}{RT}\right)\right]C_A C_B \qquad (6.71)$$

Energy balance for the reactor (J):

$$\frac{dT}{dt} = \frac{Q}{V}(T_{in} - T) + \frac{UA}{V\rho c_p}(T_j - T) + \frac{\left[k_0 \exp\left(-\frac{E_A}{RT}\right)\right]C_A C_B(-\Delta H_R)}{\rho c_p} \qquad (6.72)$$

Energy for the cooling fluid (J):

$$\frac{dTj}{dt} = \frac{Q_j}{V_j}(Tj_{in} - Tj) + \frac{UA}{V_j \rho_j c_{pj}}(T - Tj) \qquad (6.73)$$

The parameters of the model and the initial conditions are the ones presented in Tables 3.1, 3.2 and 3.3, but assuming that the transient regime started because the flow rate (Q), which was 3 m³/min (see Table 3.2), has now changed to 4 m³/min. It was also said in Example 3.8 that the volume for the jacket (V_j) is 0.032 m³.

To solve this system of ODEs (Eqs. 6.69, 6.70, 6.71, 6.72, 6.73), the codes in Fig. 6.15b (for Euler) or Fig. 6.16 (for RK4) have to be changed to include the system with five ODEs. Here we introduce the concept of an *array*, because there are five dependent variables (y) in our equations system. Using an array, we can keep the same name for the dependent variable (y) and use a number (index) to call them apart. The individual values are called the *elements of the array*. In our case we can consider:

$$y(1) = C_A$$
$$y(2) = C_B$$
$$y(3) = C_C$$
$$y(4) = T$$
$$y(5) = Tj$$

In the same way, we will also assume the variable *dydx* as an array. So, for our case it yields:

$$dydx(1) = dC_A/dt$$
$$dydx(2) = dC_B/dt$$
$$dydx(3) = dC_C/dt$$
$$dydx(4) = dT/dt$$
$$dydx(5) = dTj/dt$$

An example of how the function *Derivative* could be written to represent Eqs. (6.69), (6.70), (6.71), (6.72), (6.73) is presented in Fig. 6.17. The code of this function can be applied for both Euler and RK4 methods.

```
Function Derivative(x, y, dydx)

    Q = 4                    ' (m3/mim)
    V = 40                   ' (m3)
    Cain = 200               ' (mol/m3)
    Cbin = 200               ' (mol/m3)
    Ccin = 0                 ' (mol/m3)
    ko = 820000              ' (m3/mol min)
    Ea = 48500               ' (J/mol)
    R = 8.314                ' (J/molK)
    U = 680                  ' (J/minm2K)
    A = 5                    ' (m2)
    rho = 880                ' (kg/m3)
    Cp = 1750                ' (J/kg K)
    rhoj = 1000              ' (kg/m3)
    Cpj = 4180               ' (J/kg K)
    DeltaHr = -72800         ' (J/mol)
    Qj = 0.01                ' (m3/mim)
    Vj = 0.032               ' (m3)
    Tin = 300                ' (K)
    Tjin = 280               ' (K)

    dydx(1) = (Q / V) * (Cain - y(1)) - ko * Exp(-Ea / (R * y(4))) * y(1) * y(2)
    dydx(2) = (Q / V) * (Cbin - y(2)) - ko * Exp(-Ea / (R * y(4))) * y(1) * y(2)
    dydx(3) = (Q / V) * (Ccin - y(3)) + ko * Exp(-Ea / (R * y(4))) * y(1) * y(2)
    dydx(4) = (Q / V) * (Tin - y(4)) + (U * A / (rho * V * Cp)) * (y(5) - y(4)) _
                      + (1 / (rho * Cp)) * ko * Exp(-Ea / (R * y(4))) * y(1) * y(2) * (-DeltaHr)
    dydx(5) = (Qj / Vj) * (Tjin - y(5)) + (U * A / (rhoj * Vj * Cpj)) * (y(4) - y(5))

End Function
```

Fig. 6.17 Function *Derivative* representing the system of ordinary differential equations (ODEs) (Eqs. 6.69, 6.70, 6.71, 6.72, 6.73)

In the first lines of the code in Fig. 6.17 appear the parameters from Tables 3.1 and 3.2. These parameters could also be read from the spreadsheet by using the notation $Q = Cells(1,1)$, for example, if one wants to read the variable Q in the cell $A1$ of the spreadsheet. Equation (6.72) is very long, so it has been broken into two lines, using the underscore symbol (_).

The main programs to solve Eqs. (6.69), (6.70), (6.71), (6.72), (6.73) for the Euler and RK4 methods are shown in Fig. 6.18a and 6.18b, respectively. Observe that they are the same, only the name of the function called is different (*RungeKutta1* or *RungeKutta4*). Compare them with the main program in Fig. 6.15b. The first difference is that the dimension (*Dim*) of the array for y and *dydx* (in our case, the size is 5) must be declared. You can choose to store any number, or array of numbers, as single or double precision, but in our case, we choose double precision. Also observe that we need a loop to print all dependent variables y (from 1 to 5).

A suggestion of how the functions *RungeKutta1* and *RungeKutta4* (called in the codes of Figs. 6.18a and 6.18b) could be written is shown in Figs. 6.19 and 6.20.

For the Euler method (*RungeKutta1*), compare Figs. 6.19 and 6.15b. When solving the system of ODEs, the derivative $k1$ has dimension 5 (Eqs. 6.69, 6.70, 6.71, 6.72, 6.73), so $k1$ is declared at the beginning of the function *RungeKutta1* code as double precision array (see Fig. 6.19). Also observe in Fig. 6.19 the loop needed to calculate $k1$ and y, in order to take into account the five ODEs.

For the RK4 method, compare Figs. 6.20 (a system of ODEs) and 6.16 (just one ODE). The first difference between them is that variables $k1, k2, k3, k4$ and *ytran* are arrays with dimensions equal to five, to account for all ODEs (Eqs. 6.69, 6.70, 6.71, 6.72, 6.73), and must be declared at the beginning of the code (see Fig. 6.20). For

(General)

```
Sub Euler ()

Dim dydx(5) As Double
Dim y(5) As Double

xi = 0
xf = 25
y(1) = 49.5
y(2) = 49.5
y(3) = 150.5
y(4) = 307
y(5) = 282

dx = 0.125

x = xi

Cells(1, 1) = "x"
Cells(1, 2) = "Y1"
Cells(1, 3) = "Y2"
Cells(1, 4) = "Y3"
Cells(1, 5) = "Y4"
Cells(1, 6) = "Y5"

i = 2

ny = 5

Do While x <= xf

    Cells(i, 1) = x
    For k = 1 To ny
        Cells(i, k + 1) = y(k)
    Next

    Call RungeKutta1(x, y, dydx, ny, dx)

    i = i + 1

Loop

End Sub
```

a

```
Sub RK4 ()

Dim dydx(5) As Double
Dim y(5) As Double

xi = 0
xf = 25
y(1) = 49.5
y(2) = 49.5
y(3) = 150.5
y(4) = 307
y(5) = 282

dx = 0.125

x = xi

Cells(1, 1) = "x"
Cells(1, 2) = "Y1"
Cells(1, 3) = "Y2"
Cells(1, 4) = "Y3"
Cells(1, 5) = "Y4"
Cells(1, 6) = "Y5"

i = 2

ny = 5

Do While x <= xf

    Cells(i, 1) = x
    For k = 1 To ny
        Cells(i, k + 1) = y(k)
    Next

    Call RungeKutta4(x, y, dydx, ny, dx)

    i = i + 1

Loop

End Sub
```

b

Fig. 6.18 Examples of the main program used to solve a system of ordinary differential equations (ODEs) using **a** the Euler method and **b** the fourth-order Runge–Kutta (RK4) method

```
Function RungeKutta1(x, y, dydx, ny, dx)

Dim k1(5) As Double

'Calculate K1 for all ODEs and update dependent varables
Call Derivative(x, y, dydx)
For i = 1 To ny
    k1(i) = dydx(i)
    y(i) = y(i) + k1(i) * dx
Next

'Update the indenpendent variable
x = x + dx

End Function
```

Fig. 6.19 Example of a function for the Euler method to solve a system of ordinary differential equations (ODEs)

```
Function RungeKutta4(x, y, dydx, ny, dx)|

Dim k1(5) As Double
Dim k2(5) As Double
Dim k3(5) As Double
Dim k4(5) As Double
Dim ytran(5) As Double

'Calculate K1 for all ODEs
Call Derivative(x, y, dydx)
For i = 1 To ny
    k1(i) = dydx(i)
Next
'Define x and y to calculate K2
For i = 1 To ny
    ytran(i) = y(i) + k1(i) * 0.5 * dx
Next
xtran = x + 0.5 * dx

'Calculate K2 for all ODEs
Call Derivative(xtran, ytran, dydx)
For i = 1 To ny
    k2(i) = dydx(i)
Next
'Define x and y to calculate K3
For i = 1 To ny
    ytran(i) = y(i) + k2(i) * 0.5 * dx
Next

'Calculate K3 for all ODEs
Call Derivative(xtran, ytran, dydx)
For i = 1 To ny
    k3(i) = dydx(i)
Next
'Define x and y to calculate K4
For i = 1 To ny
    ytran(i) = y(i) + k3(i) * dx
Next
xtran = x + dx

'Calculate K4 for all ODEs
Call Derivative(xtran, ytran, dydx)
For i = 1 To ny
    k4(i) = dydx(i)
Next

'Uptade dependent variables
For i = 1 To ny
    y(i) = y(i) + (1 / 6) * (k1(i) + 2 * k2(i) + 2 * k3(i) + k4(i)) * dx
Next

'Update the independent variable
x = x + dx

End Function
```

Fig. 6.20 Example of a function for a fourth-order Runge–Kutta method to solve a system of ordinary differential equations (ODEs)

the same reason, loops are needed to calculate $k1, k2, k3, k4, y$ and $ytran$ for all ODEs (observe in Fig. 6.20 the loops going from 1 to ny).

In Figs. 6.19 and 6.20, the numbers of ODEs (ny) and $dydx$ are also passed as arguments, because ny and the dimension of $dydx$ are defined in the main program.

No matter which system of ODEs you need to solve, the Functions presented in Fig. 6.19 and 6.20 can be used as they are for Euler and RK4 methods. The only thing to change is to adjust the dimension for $k1, k2, k3, k4$ and $ytran$ if your system has a number of equations different from 5.

All codes presented in this chapter are just suggestions of how a program could be developed to solve a system of ODEs by Runge–Kutta methods using Visual Basic. Readers can find their own style of programing.

Proposed Problems
6.1) Imagine the three interconnected tanks studied in Example 5.1. Assume that the volume of liquid in the three tanks is the same and remains constant and equal to V (m³). The volumetric flow rates for all tanks are the ones presented in Table 5.1. At the beginning the three tanks contain pure water, but at a certain point, the streams Q_{01} and Q_{02} start feeding tanks 1 and 2 with a NaOH solution with concentrations of 10 mol/m³ (C_{01}) and 1 mol/m³ (C_{02}), respectively, instead of pure water, at the same flow rates. The system of ODEs that represents the variation in the concentration of NaOH over time in the three tanks is presented below:

$$\frac{dC_1}{dt} = \frac{1}{V}(50 - 7C_1 + 2C_3) \quad \text{at } t = 0, C_1 = 0$$

$$\frac{dC_2}{dt} = \frac{1}{V}(7C_1 + 1 - 8C_2) \quad \text{at } t = 0, C_2 = 0$$

$$\frac{dC_3}{dt} = \frac{1}{V}(8C_2 - 8C_3) \quad \text{at } t = 0, C_3 = 0$$

a. Solve the ODE system using the Euler method and an Excel spreadsheet, as per Sect. 6.3. Assume initially that $V = 5\text{m}^3$ and an increment (step) equal to 0.2 min. Plot curves of the concentration of NaOH for each tank over time and compare them with the concentrations in a steady state obtained in Example 5.1. Find the ideal step for this operating condition and numerical method.

b. Change the volume of the three tanks from 5 m³ to 1 m³ and check what occurs with the three curves. Keep the volume for the three tanks equal to 1 m³ but use an increment (step) for the Euler method equal to 0.1 min and observe the curves. What can you conclude?

c. Change the volume of the three tanks to 8 m³ and then to 2 m³ and observe the time needed to achieve a steady state. Are the concentrations in the steady state the same? Why?

d. Alter the initial concentrations for all tanks to 2 mol/m³. Are the concentrations in the steady state the same? Why?

e. Develop a code in Visual Basic and solve the system of three ODEs using the Euler method. Assume that $V = 5$ m^3 for the three tanks with an increment (step) equal to 0.2 min.

f. Develop a code in Visual Basic and solve the system of three ODEs using the RK4 method. Also assume that $V = 5$ m^3 for the three tanks with an increment (step) equal to 0.2 min. Compare the results with the one obtained in item (e).

g. Use the code developed in item (f) and run your program considering $V = 1$ m^3 and a step equal to 0.2 min (as was done in item (b)). Compare the results with the ones you obtained using the Euler method (in item (b)). What can you conclude?

6.2) Consider the four insulated CSTRs presented in Proposed Problem 5.2. Find the system of four ODEs that represent this reaction system. Assume that initially all reactors have concentration of reactant A equal to 1 mol/l.

a. Develop a code in Visual Basic using the Euler method, find the profiles of the concentration of A over time, and compare the results with the one obtained in Proposed Problem 5.2 for a steady state.

b. Repeat item (a), but this time use the RK4 method

c. Double the concentration of A at the initial condition ($t = 0$) for the four reactors and observe the concentration of A in a steady state. Make comments.

d. Imagine that after reaching a steady state, the concentration of A fed into the first reactor ($C_{A_{in}}$) is doubled. Find the new steady state. Make comments.

e. Repeat item (a) or (b) but this time double all kinetic constants and check what happens.

6.3) Tubular chemical reactors are widely used in the chemical industry. To develop mathematical models to describe them, it is common to assume their operations with no radial gradients of temperature, velocity, or concentration. In this case, we have plug-flow reactors (PFRs). Consider the following irreversible reaction in the plug-flow reactor:

$$A + B \rightarrow C$$

The rate equation is elementary and the reaction is carried out isothermally at 300 K in a PFR in a steady state. The feed stream has a volumetric flow rate of $Q = 10$ L/min and has both reactants, A and B, with concentrations of $C_{A0} = 1$ M and $C_{B0} = 2$ M, respectively (A and B do not react before entering the PFR). At 300 K, the rate constant (k) is 0.07 L/mol min.

a. Write ODEs to represent the concentration of A, B, and C along the reactor length.

b. Solve the ODE system using VBA and RK4. Plot the concentration profiles for A, B, and C along the reactor length. Determine at what length the conversion reaches 90% (Hint: conversion is always calculated using the limiting reactant as the reference). Choose two different values for the radius of the PFR. Plot the curves and compare the results.

6.4) Consider a CSTR in a steady state in which the irreversible and isothermal reaction $A \xrightarrow{k} B$ takes place, with a rate constant (k) equal to 10 min^{-1}. The reactor is fed with a solution of reactant A in a flow rate (Q) of 5 m^3/min and a concentration of A ($C_{A_{in}}$) of 1400 mol/m^3. The same flow rate leaves the reactor and the density of the solution does not change, so the reaction volume is constant over the reaction.

a. Develop mathematical models to represent the mass balance of A in four different situations: (i) assuming just one CSTR with a reaction volume equal to 10 m^3; (ii) assuming that the reaction system is composed of two CSTRs of 5 m^3 each in series; (iii) assuming five CSTRs of 2 m^3 each in series; (iv) assuming ten CSTRs of 1 m^3 each in series.
b. Solve the algebraic equations obtained in item (a) using a numerical method presented in Chap. 5 and compare the final concentration of reactant A leaving the reaction system for the four cases. What can you conclude?
c. Imagine there is available a PFR with a cross-sectional area and length equal to 1 m^2 and 10 m, respectively, making the reaction volume also equal to 10 m^3. Assume a steady state and the same flow rate ($Q = 5$ m^3/min) and reactant concentration ($C_{A_{in}} = 1400$ mol/m^3) feeding the system. Develop a mathematical model to represent the concentration of A along the PFR.
d. Solve the mathematical model generated in item (c) using a numerical method presented in this chapter and obtain the profile of the reactant concentration along the PFR length.
e. Compare the results obtained in items (b) and (d) by plotting, in the same graph, the concentration of A versus the reactor volume. What can you conclude?

6.5) This system is adapted from Incropera et al. (2006). A very long cylindrical metal bar with diameter D, length L, and thermal conductivity k has one end maintained at T_w by constant contact with a hot wall. The surface of this cylinder is exposed to ambient air at a constant temperature of T_∞ with a convection heat transfer coefficient of h. The system was left for a long time until it became completely stable.

a. Write ODEs that describes the temperature profile and define the two boundary conditions. Consider that there is no radial temperature profile inside the bar. Make assumptions if needed to simplify the mathematical solution of this problem.
b. Solve the mathematical model using VBA and RK4.
    Determine and plot the temperature profiles along the bar length when it is manufactured from pure copper, aluminum, and stainless steel.
    Consider the following numerical values:

$$D = 5 \text{ mm}$$
$$T_w = 100\,^\circ C$$
$$T_\infty = 25\,^\circ C$$
$$h = 100W/(m^2K)$$

$$\text{Copper:} \quad k = 398 \text{W}/(\text{m K})$$
$$\text{Aluminum:} \quad k = 180 \text{W}/(\text{m K})$$
$$\text{Stainless steel:} \quad k = 14 \text{W}/(\text{m K})$$

c. Determine for each metallic material the minimum length that the bar must have for the bar temperature profile to reach a minimum plateau. Using the minimum lengths for each one of the metals, determine the heat loss for each material.

Hint: In order to numerically integrate second-order ODEs more easily, the following substitution can be very handy:

$$\frac{dT}{dx} = y = f(x)$$
$$\frac{d^2T}{dx^2} = \frac{dy}{dx} = g(x)$$

Instead of directly solving one single second-order differential equation, it is possible to break it into two first-order differential equations to be solved independently. Observe that, in this problem, one of the boundary conditions must suffer this change in variable too.

6.6) Imagine the two concentric cylinders modeled in Example 4.8 and assume that the system has reached a steady state, so it can be represented by:

$$r\frac{d^2T}{dr^2} + \frac{dT}{dr} = 0$$
$$\text{At } r = R_1, \quad T = T_0$$
$$\text{At } r = R_2, \quad \frac{dT}{dr} = -\frac{h}{k}(T - T_{env})$$

Assume the following numerical values: $R_1 = 0.5$ cm, $R_2 = 3$ cm, $T_0 = 100\,°\text{C}$, $T_{env} = 25\,°\text{C}, k = 180 \text{W}/(\text{mK})$, and $h = 100\ W/(m^2 K)$. Use the same hint suggested in Proposed Problem 6.5 and solve this problem using VBA and RK4. Plot the radial profile of the temperature.

References

Billo, E.J.: Excel for Scientists and Engineers Numerical Methods. Wiley, Hoboken (2007)
Chapra, C.C., Canale, R.P.: Numerical Methods for Engineers, 5th edn. McGraw Hill, New York (2005)
Davis, M.E.: Numerical Methods and Modeling for Chemical Engineers. Wiley, New York (1984)
Incropera, F.P., DeWitt, D.P., Bergman, T.L., Lavine, A.S.: Introduction to Heat Transfer, 5th edn. Wiley, Hoboken (2006)
Rao, S.S.: Applied Numerical Methods for Engineers and Scientists. Prentice Hall, Upper Saddle River (2002)
Varma, A., Morbidelli, M.: Mathematical Methods in Chemical Engineering. Oxford University Press, Oxford (1997)
Walkenbach, J.: Excel VBA Programing for Dummies, 3rd edn. Wiley, Hoboken (2013a)
Walkenbach, J.: Excel Bible. Wiley, Hoboken (2013b)

Chapter 7
Solving a Partial Differential Equations System

The idea for solving a system of partial differential equations (PDEs) using numerical methods is to transform it into a system of equations that are easier to solve, such as algebraic equations or ordinary differential equations (ODEs), for which numerical solutions were presented in Chaps. 5 and 6 of this book.

There are many numerical methods to solve PDEs, such as finite difference, finite volume, orthogonal collocation, etc., but this book will focus on the finite difference method. Other numerical methods can be found elsewhere in the literature (Davis 1984; Chapra and Canale 2005, etc.).

7.1 Motivation

Consider the insulated cylindrical metal bar of 1 m studied in Example 4.4. Initially this bar is at 50 °C, but it is fixed between two walls at temperatures of 70 °C and 30 °C, as depicted in Fig. 4.9. The modeling of this system generates the following PDE and initial/boundary conditions (see details in Chap. 4):

$$\frac{\partial T}{\partial t} = \frac{k}{\rho c_p} \frac{\partial^2 T}{\partial x^2} \tag{4.15}$$

At $t = 0$ h, $T = 50$ °C, for $0 \le L \le 1$ m
At $x = 0$ m, $T = 70$ °C, for $t > 0$ h
At $x = 1$ m, $T = 30$ °C, for $t > 0$ h

The finite difference method will represent the partial derivative of temperature with respect to length $(\partial^2 T/\partial x^2)$ or both derivatives $(\partial^2 T/\partial x^2$ and $\partial T/\partial t)$ by

---

The original version of this chapter was revised. An erratum to this chapter can be found at
https://doi.org/10.1007/978-3-319-66047-9_8

expressions that are easier to solve. Section 7.2 presents the finite difference method, and Sects. 7.3 and 7.4 show how this numerical method can be used to solve Eq. (4.15) and other PDEs.

7.2   Finite Difference Method

Imagine a generic dependent variable u (which could be temperature, concentration, etc.) changing with two generic independent variables x and y (such as time, length, radius, etc.). The indices related to x and y will be called i and j, respectively. Knowing the value of u in a certain condition of x and y ($u_{i,j}$), it is possible to represent u after an infinitesimal increment of x (Δx) or y (Δy) using expansion of the Taylor series. In order to do that, let us imagine the grid denoted by Fig. 7.1. The idea of the finite difference method is to approximate the values of the continuous function u by a set of discrete points in the (x, y) plane, which we call *discretization*.

Given $u_{i,j}$, it is possible to obtain $u_{i+1,j}$ and $u_{i-1,j}$ using expansion of the Taylor series (see also Eq. 6.3), as shown below:

$$u_{i+1,j} = u_{i,j} + \Delta x \left(\frac{\partial u}{\partial x}\right)_{i,j} + \frac{1}{2!}\Delta x^2 \left(\frac{\partial^2 u}{\partial x^2}\right)_{i,j} + \cdots + \frac{1}{n!}\Delta x^n \left(\frac{\partial^n u}{\partial x^n}\right)_{i,j} + Rn \quad (7.1)$$

$$u_{i-1,j} = u_{i,j} + (-\Delta x)\left(\frac{\partial u}{\partial x}\right)_{i,j} + \frac{1}{2!}(-\Delta x)^2 \left(\frac{\partial^2 u}{\partial x^2}\right)_{i,j} + \cdots + \frac{1}{n!}(-\Delta x)^n \left(\frac{\partial^n u}{\partial x^n}\right)_{i,j} + Rn$$

$$(7.2)$$

Remember that in the finite difference method we want to find simpler expressions for the second and/or first partial derivatives. So if Eqs. (7.1) and (7.2) are truncated after the first term, it is possible to obtain expressions for the first derivative, as shown below:

Fig. 7.1 Grid to visualize the finite difference method

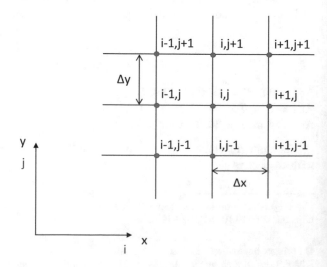

From Equation (7.1):
$$\left(\frac{\partial u}{\partial x}\right)_{i,j} = \frac{u_{i+1,j} - u_{i,j}}{\Delta x} + R_1 \qquad (7.3)$$

From Equation (7.2):
$$\left(\frac{\partial u}{\partial x}\right)_{i,j} = \frac{u_{i,j} - u_{i-1,j}}{\Delta x} + R_1 \qquad (7.4)$$

Equations (7.3) and (7.4) are called *forward difference* and *backward difference*, respectively, and both are *one-sided difference* expressions. The first term after truncation of the Taylor series contains the major error, and for both expressions, this error is proportional to $(\Delta x)^2$ (see Eqs. 7.1 and 7.2).

A more accurate way to represent the first derivative can be obtained by subtracting Eq. (7.2) from Eq. (7.1), both truncated after the second term, which yields Eq. (7.5), which is called a *centered difference* formula.

Equation (7.1) minus (7.2):
$$\left(\frac{\partial u}{\partial x}\right)_{i,j} = \frac{u_{i+1,j} - u_{i-1,j}}{2(\Delta x)} + R_2 \qquad (7.5)$$

Observe that the first term not considered in Eq. (7.5) is proportional to $(\Delta x)^3$, because the terms with $(\Delta x)^2$ are naturally cancelled. Since Δx is very small, $(\Delta x)^3 < (\Delta x)^2$, which guarantees a smaller error for the *centered difference* formula.

The second partial derivative could be obtained by adding Eqs. (7.1) and (7.2) truncated after the third term:

Adding Eqs. (7.1) and (7.2):
$$\left(\frac{\partial^2 u}{\partial x^2}\right)_{i,j} = \frac{u_{i-1,j} - 2u_{i,j} + u_{i+1,j}}{(\Delta x)^2} + R_3 \qquad (7.6)$$

In the same way, given $u_{i,j}$, it is possible to obtain $u_{i,j+1}$ and $u_{i,j-1}$ using expansion of the Taylor series, generating Eqs. (7.7) and (7.8):

$$u_{i,j+1} = u_{i,j} + \Delta y \left(\frac{\partial u}{\partial y}\right)_{i,j} + \frac{1}{2!}\Delta y^2 \left(\frac{\partial^2 u}{\partial y^2}\right)_{i,j} + \cdots + \frac{1}{n!}\Delta y^n \left(\frac{\partial^n u}{\partial y^n}\right)_{i,j} + Rn \quad (7.7)$$

$$u_{i,j-1} = u_{i,j} + (-\Delta y)\left(\frac{\partial u}{\partial y}\right)_{i,j} + \frac{1}{2!}(-\Delta y)^2\left(\frac{\partial^2 u}{\partial y^2}\right)_{i,j} + \cdots + \frac{1}{n!}(-\Delta y)^n\left(\frac{\partial^n u}{\partial y^n}\right)_{i,j} + Rn$$
$$(7.8)$$

If Eqs. (7.7) and (7.8) are truncated after the first term, it is possible to obtain the *one-sided difference* expressions for the first derivative of u with respect to y:

From Eq. (7.7):
$$\left(\frac{\partial u}{\partial y}\right)_{i,j} = \frac{u_{i,j+1} - u_{i,j}}{\Delta y} + R_1 \qquad (7.9)$$

From Eq. (7.8):
$$\left(\frac{\partial u}{\partial y}\right)_{i,j} = \frac{u_{i,j} - u_{i,j-1}}{\Delta y} + R_1 \qquad (7.10)$$

A *centered difference* formula is also obtained by subtracting Eq. (7.8) from Eq. (7.7), both truncated after the second term:

$$\text{Eq. (7.7) minus Eq. (7.8):} \quad \left(\frac{\partial u}{\partial y}\right)_{i,j} = \frac{u_{i,j+1} - u_{i,j-1}}{2(\Delta y)} + R_2 \qquad (7.11)$$

The second partial derivative of u with respect to y can be obtained by adding Eqs. (7.7) and (7.8) truncated after the third term:

$$\text{Adding Eqs. (7.7) and (7.8):} \quad \left(\frac{\partial^2 u}{\partial y^2}\right)_{i,j} = \frac{u_{i,j-1} - 2u_{i,j} + u_{i,j+1}}{(\Delta y)^2} + R_3 \qquad (7.12)$$

Now that we have obtained expressions for the first and second derivatives, let us apply them in practical examples, as depicted in Sects. 7.3 and 7.4.

7.3   Introductory Example of Finite Difference Method Application

In this section, we will apply the expressions obtained in Sect. 7.2 for the first and second derivatives to solve Eq. (4.15) in Sect. 7.1.

One way to numerically solve Eq. (4.15) (rewritten below) is to discretize all partial derivatives:

$$\frac{\partial T}{\partial t} = \frac{k}{\rho c_p} \frac{\partial^2 T}{\partial x^2} \qquad (4.15)$$

Equation (7.13) can be used to represent the second-order derivative of T with respect to x, (compare with Eq. 7.6).

$$\left(\frac{\partial^2 T}{\partial x^2}\right)_{i,j} = \frac{T_{i-1,j} - 2T_{i,j} + T_{i+1,j}}{(\Delta x)^2} + R_3 \qquad (7.13)$$

where i and j represent x and t, respectively.

In theory, Eqs. (7.9), (7.10) and (7.11) show three options to represent the first derivative with respect to the other independent variable (t). We will use Eq. (7.9) and explain why later on.

$$\left(\frac{\partial T}{\partial t}\right)_{i,j} = \frac{T_{i,j+1} - T_{i,j}}{\Delta t} + R_1 \qquad (7.14)$$

Substituting Eqs. (7.13) and (7.14) in Eq. (4.15) and neglecting the truncation errors we obtain:

$$\frac{T_{i,j+1} - T_{i,j}}{\Delta t} = \frac{k}{\rho c_p} \left(\frac{T_{i-1,j} - 2T_{i,j} + T_{i+1,j}}{(\Delta x)^2} \right) \tag{7.15}$$

Equation 7.15 can be rearranged to yield:

$$T_{i,j+1} = Fo \left(T_{i-1,j} - 2T_{i,j} + T_{i+1,j} \right) + T_{i,j} \tag{7.16}$$

in which:

$$Fo = \frac{k}{\rho c_p} \frac{\Delta t}{(\Delta x)^2} \tag{7.17}$$

The numerical values for k, c_p, and ρ are the ones presented in Example 4.4: $k = 398.2$ J/s m °C (or $k = 23{,}892$ J/min m °C), $c_p = 386.3$ J/kg °C, and $\rho = 8933$ kg/m^3.

The lower the values of Δx and Δt are, the closer the numerical solution will be to the analytical results, which we call *convergence*. *Stability* means no propagation of errors as the numerical calculations are done. For this problem, the method is stable and convergent if Fo in Eq. (7.17) is lower than 0.5 (Carnahan et al. 1969). In this problem, we will assume that $\Delta x = 0.2$ m and $\Delta t = 0.05$ min, which guarantees a safe, stable, and convergent solution.

Based on Eq. (7.16), it is possible to build a table in Excel (see Fig. 7.2), in which the temperature of the bar at $t = 0$ min (50 °C) and at $x = 0$ (70 °C) and $x = 1$ m (30 °C) can be written. If the temperature at $T_{i,j+1}$ is the temperature at $x = 0.2$ m and $t = 0.05$ min (see point highlighted in Fig. 7.2), $T_{i-1,j}$, $T_{i,j}$, and $T_{i+1,j}$ are the temperatures at time (j) equal to zero and length (i) equal to 0, 0.2, and 0.4, respectively. Since $T_{i-1,j}$, $T_{i,j}$, and $T_{i+1,j}$ are known, it is possible to obtain $T_{i,j+1}$ from Eq. (7.16), as depicted by the function at the top of the spreadsheet. Temperatures for other values of length and time can be obtained by dragging the *Fill Handle* down and right.

Observe that, in this case, $\partial T/\partial t$ could not be represented in the more accurate way (Eq. 7.11—the *centered difference* formula), because to obtain $T_{i,j+1}$ the value at $T_{i,j-1}$ (the time before 0 min) would be needed (see Eq. 7.11).

Using the results shown in Fig. 7.2, it is possible to obtain the axial profiles of temperature over time. If you plot the results, you will see that the profiles of temperature are similar to the ones shown in Fig. 4.6b.

7.4 Application of the Finite Difference Method

This section presents the application of finite difference in four different situations. Section 7.4.1 shows an example in which a system of algebraic equations is generated after discretization in all independent variables. In fact, PDEs can be

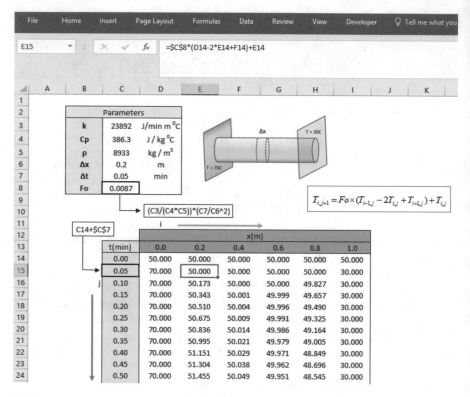

Fig. 7.2 Solving a partial differential equation (PDE) (4.15) using the finite difference method and Excel

solved considering discretization in all independent variables (as in the previous example and Sect. 7.4.1) or keeping the derivative in one independent variable, which generates a system of ordinary differential equations. Section 7.4.2 will revisit the previous example (shown in Sect. 7.3) but will perform the discretization only in the space coordinate, which generates an ODE system varying over time. Section 7.4.3 will apply the finite difference method to a system of PDEs (not to just one equation). Finally, Sect. 7.4.4 will study PDEs with flux boundary conditions.

7.4.1 PDEs Transformed into an Algebraic Equations System

In the example presented in Sect. 7.3, the discretization in all independent variables resulted in a single algebraic Eq. (7.16), whose solution for different values of i (length) and j (time), in a sequential way, can give us axial profiles of temperature over time. Sometimes, discretization in all independent variables generates a system of many algebraic equations that must be solved simultaneously. The next example will address this situation.

Modeling of the heat conduction along a square metal plate with negligible thickness generates the following PDE:

$$\frac{\partial^2 T}{\partial x^2} + \frac{\partial^2 T}{\partial y^2} = 0 \qquad (7.18)$$

In this case, we assume a steady state and no heat exchange with the environment. The temperatures at the plate ends are fixed, as shown in Fig. 7.3a. This system is also studied in Chapra and Canale (2005).

Discretizing both derivatives of Eq. (7.18) and assuming truncation errors negligible yields (see Eqs. 7.6 and 7.12):

$$\frac{T_{i-1,j} - 2T_{i,j} + T_{i+1,j}}{(\Delta x)^2} + \frac{T_{i,j-1} - 2T_{i,j} + T_{i,j+1}}{(\Delta y)^2} = 0 \qquad (7.19)$$

Assuming $\Delta x \cong \Delta y$ and rearranging we obtain:

$$T_{i+1,j} + T_{i-1,j} + T_{i,j+1} + T_{i,j-1} - 4T_{i,j} = 0 \qquad (7.20)$$

In this example, we will divide the plate into 16 equal parts, as shown in the grid in Fig. 7.3b.

Figure 7.3b depicts that, besides the points at the edges (known temperatures), there are nine points inside the grid at which the temperatures are unknown. To obtain the temperatures at these nine points, we apply Eq. (7.20) assuming i and j varying from 1 to 3 (see Fig. 7.3b) to obtain the linear algebraic equations system shown in Table 7.1.

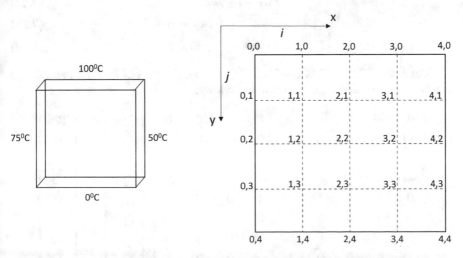

Fig. 7.3 (a) Square metal plate showing the temperatures at the ends. (b) Grid with nine unknown internal temperature points

Table 7.1 Linear algebraic equations system representing the temperatures inside the plate

i	j	(i,j)	Equation
1	1	(1,1)	$T_{2,1}+T_{0,1}+T_{1,2}+T_{1,0}-4T_{1,1}=0$
2	1	(2,1)	$T_{3,1}+T_{1,1}+T_{2,2}+T_{2,0}-4T_{2,1}=0$
3	1	(3,1)	$T_{4,1}+T_{2,1}+T_{3,2}+T_{3,0}-4T_{3,1}=0$
1	2	(1,2)	$T_{2,2}+T_{0,2}+T_{1,3}+T_{1,1}-4T_{1,2}=0$
2	2	(2,2)	$T_{3,2}+T_{1,2}+T_{2,3}+T_{2,1}-4T_{2,2}=0$
3	2	(3,2)	$T_{4,2}+T_{2,2}+T_{3,3}+T_{3,1}-4T_{3,2}=0$
1	3	(1,3)	$T_{2,3}+T_{0,3}+T_{1,4}+T_{1,2}-4T_{1,3}=0$
2	3	(2,3)	$T_{3,3}+T_{1,3}+T_{2,4}+T_{2,2}-4T_{2,3}=0$
3	3	(3,3)	$T_{4,3}+T_{2,3}+T_{3,4}+T_{3,2}-4T_{3,3}=0$

The equations in Table 7.1 can be solved using the procedure presented in Sect. 5.1 of this book. The temperatures in the equations in Table 7.1 that presents i or j equal to 0 or 4 are known (see Fig. 7.3) because they belong to the edges. Substituting these known values in the equations in Table 7.1 and rearranging, we obtain the system of algebraic equations presented by Eqs. (7.21), (7.22), (7.23), (7.24), (7.25), (7.26), (7.27), (7.28) and (7.29).

$$4T_{1,1} - T_{1,2} - T_{2,1} = 175 \tag{7.21}$$

$$-T_{1,1} + 4T_{2,1} - T_{2,2} - T_{3,1} = 100 \tag{7.22}$$

$$-T_{2,1} + 4T_{3,1} - T_{3,2} = 150 \tag{7.23}$$

$$-T_{1,1} + 4T_{1,2} - T_{1,3} - T_{2,2} = 75 \tag{7.24}$$

$$T_{1,2} + T_{2,1} - 4T_{2,2} + T_{2,3} + T_{3,2} = 0 \tag{7.25}$$

$$-T_{2,2} - T_{3,1} + 4T_{3,2} - T_{3,3} = 50 \tag{7.26}$$

$$-T_{1,2} + 4T_{1,3} - T_{2,3} = 75 \tag{7.27}$$

$$T_{1,3} + T_{2,2} - 4T_{2,3} + T_{3,3} = 0 \tag{7.28}$$

$$-T_{2,3} - T_{3,2} + 4T_{3,3} = 50 \tag{7.29}$$

Equations (7.21), (7.22), (7.23), (7.24), (7.25), (7.26), (7.27), (7.28) and (7.29) can be represented in a matrix form, as follows:

$$
\begin{bmatrix}
4 & -1 & 0 & -1 & 0 & 0 & 0 & 0 & 0 \\
-1 & 0 & 0 & 4 & -1 & 0 & -1 & 0 & 0 \\
0 & 0 & 0 & -1 & 0 & 0 & 4 & -1 & 0 \\
-1 & 4 & -1 & 0 & -1 & 0 & 0 & 0 & 0 \\
0 & 1 & 0 & 1 & -4 & 1 & 0 & 1 & 0 \\
0 & 0 & 0 & 0 & -1 & 0 & -1 & 4 & -1 \\
0 & -1 & 4 & 0 & 0 & -1 & 0 & 0 & 0 \\
1 & 0 & 0 & 0 & 1 & -4 & 0 & 0 & 1 \\
0 & 0 & 0 & 0 & 0 & -1 & 0 & -1 & 4
\end{bmatrix}
\begin{Bmatrix}
T_{1,1} \\ T_{1,2} \\ T_{1,3} \\ T_{2,1} \\ T_{2,2} \\ T_{2,3} \\ T_{3,1} \\ T_{3,2} \\ T_{3,3}
\end{Bmatrix}
=
\begin{Bmatrix}
175 \\ 100 \\ 150 \\ 75 \\ 0 \\ 50 \\ 75 \\ 0 \\ 50
\end{Bmatrix}
$$

or:

$$[A]\{T\} - \{B\} \tag{7.30}$$

in which:

$[A]$ = matrix of coefficients of Eqs. (7.21), (7.22), (7.23), (7.24), (7.25), (7.26), (7.27), (7.28) and (7.29)

$\{T\}$ = vector of unknown temperatures at the nine points inside the grid

$\{B\}$ = vector of numbers related to the right side of Eqs. (7.21), (7.22), (7.23), (7.24), (7.25), (7.26), (7.27), (7.28) and (7.29)

As was done in Chap. 5, we can multiply both sides of Eq. (7.30) by the inverse of matrix A:

$$[A]^{-1}[A]\{T\} = [A]^{-1}\{B\} \tag{7.31}$$

Since $[A][A]^{-1} = [A]^{-1}[A] = [I]$, we obtain:

$$\{T\} = [A]^{-1}\{B\} \tag{7.32}$$

As was done in Chap. 5, we can build matrix A in an Excel spreadsheet, invert this matrix, and multiply the result by vector B, as depicted in Fig. 7.4.

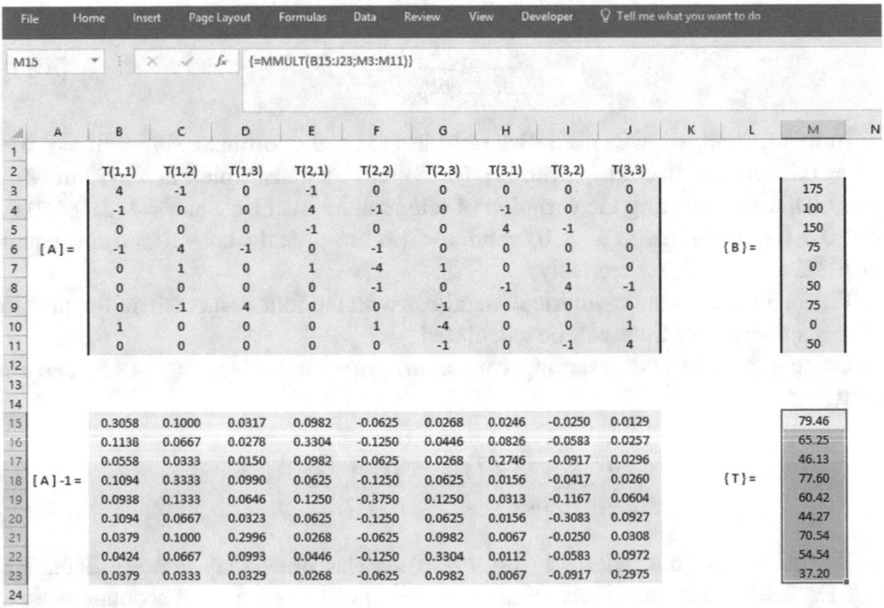

Fig. 7.4 Suggestion for an Excel spreadsheet to solve Eq. (7.30) and obtain the temperature along the plate

Fig. 7.5 Inside
temperatures of the square
plate

79.46	77.60	70.54
65.25	60.42	54.54
46.13	44.27	37.20

Figure 7.5 shows the nine temperatures inside the square plate. Observe that the points closer to the ends with higher temperatures also present higher temperatures, as expected (compare this with Fig. 7.3a).

If one wants to obtain a more precise result, more discretization points have to be considered; however, a system with more equations must be solved simultaneously.

7.4.2 PDEs Transformed into an ODE System

The problem proposed in Sect. 7.1 (Eq. 4.15, rewritten below) and solved in Sect. 7.3 could also be solved with discretization in just one independent variable.

$$\frac{\partial T}{\partial t} = \frac{k}{\rho c_p} \frac{\partial^2 T}{\partial x^2} \tag{4.15}$$

In this section, we will discretize only in the axial coordinate. We will assume $\Delta x = 0.2$, as done in Sect. 7.3, so Eq. (4.15) will be transformed in a system with four ODEs, representing the variation of temperature over time at $x = 0.2, 0.4, 0.6,$ and 0.8 (temperatures at $x = 0.0$ and $x = 1.0$ are already known and are equal to 70 °C and 30 °C, respectively).

Figure 7.6 shows the cylindrical metal bar with the four points where the profile of temperature over time will be calculated.

Using Eq. (7.6) and assuming truncation error negligible, Eq. (4.15) can be written as:

$$\frac{dT_i}{dt} = \frac{k}{\rho c_p}\left(\frac{T_{i-1} - 2T_i + T_{i+1}}{(\Delta x)^2}\right) \tag{7.33}$$

We do not use the index j in Eq. (7.33) as was previously done in Sect. 7.3 (see Eq. 7.13), because j is related to time, which will be taken into account when a numerical method to solve ODEs is used. For the different values of i, the ODE system can be written as:

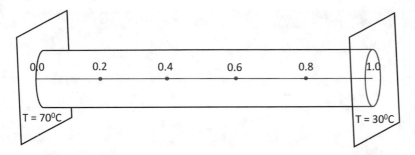

Fig. 7.6 Cylindrical bar showing the points where profiles of the temperature over time will be calculated

$$\text{At } x = 0.2: \quad \frac{dT_{0.2}}{dt} = \frac{k}{\rho c_p}\left(\frac{70 - 2T_{0.2} + T_{0.4}}{(0.2)^2}\right) \tag{7.34}$$

$$\text{At } x = 0.4: \quad \frac{dT_{0.4}}{dt} = \frac{k}{\rho c_p}\left(\frac{T_{0.2} - 2T_{0.4} + T_{0.6}}{(0.2)^2}\right) \tag{7.35}$$

$$\text{At } x = 0.6: \quad \frac{dT_{0.6}}{dt} = \frac{k}{\rho c_p}\left(\frac{T_{0.4} - 2T_{0.6} + T_{0.8}}{(0.2)^2}\right) \tag{7.36}$$

$$\text{At } x = 0.8: \quad \frac{dT_{0.8}}{dt} = \frac{k}{\rho c_p}\left(\frac{T_{0.6} - 2T_{0.8} + 30}{(0.2)^2}\right) \tag{7.37}$$

The ODE system can be solved using the numerical methods of Runge–Kutta. In this section, we will use the first-order Runge–Kutta (RK1) or Euler method (see Eq. 6.16). We can either develop a code in Visual Basic for Applications (VBA) or solve the ODE system using a spreadsheet in Excel, as was done in Sect. 6.3. We choose this second option. A suggestion for how the spreadsheet could be built is presented in Fig. 7.7. The function space in the spreadsheet shows the application of the Euler method (Eq. 6.16) to obtain the temperature at $x = 0.2$ m and $t = 0.05$ min. After typing this equation and pressing *Enter*, we can drag the *Fill Handle* right and down to complete the other values in the table. Observe that the same results are obtained in Figs. 7.2 and 7.7, as expected.

7.4.3   Solving a System of PDEs

The previous examples showed how to solve one single PDE. The methodology presented so far can also be applied to solve a system of PDEs, as can be seen in the next example adapted from Hill and Root (2014). The reaction $A \xrightarrow{k} B$, in a liquid

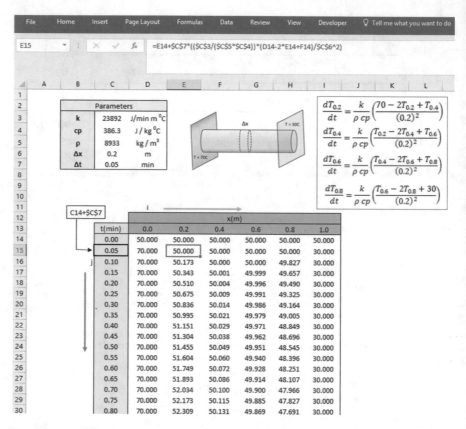

Fig. 7.7 Suggestion for a spreadsheet to solve a system of four ordinary differential equations (ODEs) using the Euler method after discretizing only in an axial coordinate

phase, occurs in a plug-flow reactor (PFR) at a constant pressure equal to 202.6 kPa. The tubular reactor is fed with a solution of reactant A in a concentration (C_{A_0}) of 18.75 kmol/m^3, a flow rate (Q) of 32 m^3/h, and a temperature (T_0) of 200 °C. The average enthalpy of the reaction (ΔH_R) is $-15{,}000$ kJ/kmol. The reaction rate constant can be represented by the expression below for the considered interval of the temperature:

$$k = 110 + 0.8 \, (T - 200) \tag{7.38}$$

in which:

$T =$ reactor temperature (°C)
$k =$ reaction rate constant (h^{-1})

The reactor operates in an adiabatic way (no heat exchange with the environment) and in a transient regime. Assume that, in the beginning (time equal to zero), the concentration of A and the temperature along the reactor are equal to 0 kmol/m^3 and 200 °C, respectively. Consider, for the reactional mixture, $\rho c_p = 787.5$ kJ/m^{3}K constant during the reaction. Assume the reactor has length (L) and diameter (D) of 3 m and 24.25 cm, respectively. We want to know the axial profiles of the temperature and the concentration of A over time.

Doing the modeling of this system according to Chap. 4, we obtain the balance of reactant A and the temperature inside the reactor as presented by Eqs. 7.39 and 7.40. No diffusion in axial and radial directions is considered.

$$\text{Balance for A:} \quad \frac{\partial C_A}{\partial t} = -\frac{Q}{A}\frac{\partial C_A}{\partial z} - kC_A \tag{7.39}$$

$$\text{Energy balance:} \quad \frac{\partial T}{\partial t} = -\frac{Q}{A}\frac{\partial T}{\partial z} + \frac{(-\Delta H_R)kC_A}{\rho c_p} \tag{7.40}$$

in which:

A = reactor cross-sectional area $=(\pi D^2)/4$

The initial/boundary conditions needed to solve this equation system are:

$$t = 0: C_A = 0, \text{ for } 0 \leq z \leq L \tag{7.41}$$

$$t = 0: T = 200°C, \text{ for } 0 \leq z \leq L \tag{7.42}$$

$$z = 0: C_A = C_{A_0} = 18.75 \text{ kmol/m}^3, \text{ for } 0 < t < \infty \tag{7.43}$$

$$z = 0: T = T_0 = 200°C, \text{ for } 0 < t < \infty \tag{7.44}$$

To solve Eqs. (7.39) and (7.40), we will consider the axial discretization in order to obtain a system of ODEs varying over time. If we assume $\Delta z = 0.6$, we can obtain an ODE system able to predict how C_A and T vary over time at z equal to 0.6 m, 1.2 m, 1.8 m, 2.4 m and 3.0 m. In theory, $\partial T/\partial z$ and $\partial C_A/\partial z$ can be discretized using *forward*, *backward*, or *centered difference* (Eqs. 7.3, 7.4, or 7.5, respectively). We will use *backward difference* (Eq. 7.4), as denoted by Eqs. (7.45) and (7.46) below. The index j, related to time, of Eq. (7.4) is not written, because time will be taken into account in the numerical method to solve the ODE system. Eqs. (7.45) and (7.46) also consider truncation error negligible.

$$\left(\frac{\partial C_A}{\partial z}\right)_i = \frac{C_{A_i} - C_{A_{i-1}}}{\Delta z} \tag{7.45}$$

$$\left(\frac{\partial T}{\partial z}\right)_i = \frac{T_i - T_{i-1}}{\Delta z} \tag{7.46}$$

Substituting (7.45) and (7.46) in Eqs. (7.39) and (7.40) yields:

$$\frac{d(C_A)_{0.6}}{dt} = -\frac{Q}{A}\frac{(C_A)_{0.6} - (C_A)_0}{\Delta z} - k_{0.6}(C_A)_{0.6} \tag{7.47}$$

$$\frac{d(C_A)_{1.2}}{dt} = -\frac{Q}{A}\frac{(C_A)_{1.2} - (C_A)_{0.6}}{\Delta z} - k_{1.2}(C_A)_{1.2} \tag{7.48}$$

$$\frac{d(C_A)_{1.8}}{dt} = -\frac{Q}{A}\frac{(C_A)_{1.8} - (C_A)_{1.2}}{\Delta z} - k_{1.8}(C_A)_{1.8} \tag{7.49}$$

$$\frac{d(C_A)_{2.4}}{dt} = -\frac{Q}{A}\frac{(C_A)_{2.4} - (C_A)_{1.8}}{\Delta z} - k_{2.4}(C_A)_{2.4} \tag{7.50}$$

$$\frac{d(C_A)_{3.0}}{dt} = -\frac{Q}{A}\frac{(C_A)_{3.0} - (C_A)_{2.4}}{\Delta z} - k_{3.0}(C_A)_{3.0} \tag{7.51}$$

$$\frac{dT_{0.6}}{dt} = -\frac{Q}{A}\frac{T_{0.6} - T_0}{\Delta z} + \frac{(-\Delta H_R)k_{0.6}(C_A)_{0.6}}{\rho c_p} \tag{7.52}$$

$$\frac{dT_{1.2}}{dt} = -\frac{Q}{A}\frac{T_{1.2} - T_{0.6}}{\Delta z} + \frac{(-\Delta H_R)k_{1.2}(C_A)_{1.2}}{\rho c_p} \tag{7.53}$$

$$\frac{dT_{1.8}}{dt} = -\frac{Q}{A}\frac{T_{1.8} - T_{1.2}}{\Delta z} + \frac{(-\Delta H_R)k_{1.8}(C_A)_{1.8}}{\rho c_p} \tag{7.54}$$

$$\frac{dT_{2.4}}{dt} = -\frac{Q}{A}\frac{T_{2.4} - T_{1.8}}{\Delta z} + \frac{(-\Delta H_R)k_{2.4}(C_A)_{2.4}}{\rho c_p} \tag{7.55}$$

$$\frac{dT_{3.0}}{dt} = -\frac{Q}{A}\frac{T_{3.0} - T_{2.4}}{\Delta z} + \frac{(-\Delta H_R)k_{3.0}(C_A)_{3.0}}{\rho c_p} \tag{7.56}$$

The concentration and temperature at $z = 0$ (C_{A_0} in Eq. 7.47 and T_0 in Eq. 7.52) are known: $C_{A_0} = 18.75$ kmol/m^3 and $T_0 = 200$, as can be seen in the boundary conditions (Eqs. 7.43 and 7.44), so there are ten unknown values and ten ODEs to be simultaneously solved. Observe that if *forward* or *centered difference* (Eqs. 7.3 or 7.5) were used to represent $\partial T/\partial z$ and $\partial C_A/\partial z$, C_A and T at $z = 3.6$ would be needed in Eqs. (7.51) and (7.56), which is longer than the total length of the reactor ($L = 3.0$ m), so *backward difference* (Eq. 7.4) was used in this example.

The system of Eqs. (7.47), (7.48), (7.49), (7.50), (7.51), (7.52), (7.53), (7.54), (7.55) and (7.56) can be solved using the Runge–Kutta methods shown in Chap. 6. For this problem, we develop a code in VBA that uses the fourth-order Runge–Kutta (RK4) method to solve the ODE system (see Appendix 7.1). The initial condition at t = 0 is given by Eqs. (7.41) and (7.42) (at $t = 0$, $C_A = 0$ and $T = 200$ °C for $0 \leq z \leq 3.0\,m$).

The program generates profiles of the concentration of reactant A and the temperature varying along the reactor length and over time, which can be visualized in Figs. 7.8 and 7.9.

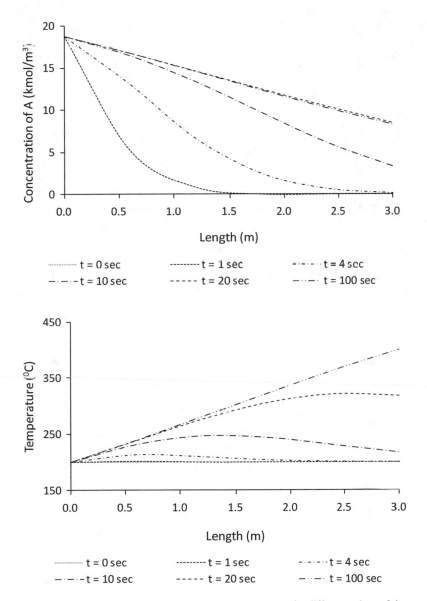

Fig. 7.8 Axial profiles of the concentration of A and temperature for different values of time until a steady state is reached

Observe that the code in VBA presented in Appendix 7.1 to generate Figs. 7.8 and 7.9 is almost the same as that used in Chap. 6, because it was developed in a generic way.

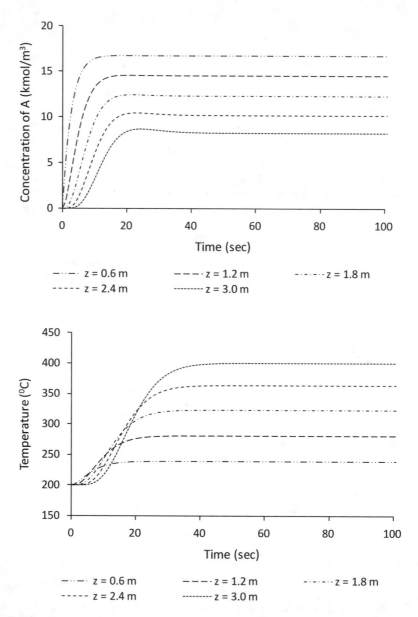

Fig. 7.9 Profiles of the concentration of A and temperature over time at different positions inside the reactor

7.4.4 PDEs with Flux Boundary Conditions

In chemical engineering, boundary conditions involving the flux of a given component occur very frequently. Imagine the insulated cylindrical metal bar considered in Sects. 7.1, 7.3, and 7.4.2 by Eq. (4.15), but, this time, assume that one of the ends exchanges heat with the environment, as depicted in Fig. 4.10 in Example 4.5. The problem to be solved is rewritten below:

$$\frac{\partial T}{\partial t} = \frac{k}{\rho c_p} \frac{\partial^2 T}{\partial x^2} \tag{4.15}$$

At $t = 0$ h, $T = 50\,°C$, for $0 \le L \le 1$ m

At $x = 0$ m, $T = 70\,°C$, for $t > 0$ h

At $x = 1m$, $\dfrac{dT}{dx} = -\dfrac{h}{k}(T - T_{env})$, for $t > 0$ h $\tag{7.57}$

in which:

k = thermal conductivity (J/s m °C)
h = coefficient of heat transfer by convection (J/s m^2 °C)
T_{env} = environment temperature = 25 °C

Equation (4.15) can be solved numerically using both approaches presented in Sects. 7.3 and 7.4.2. For both cases, the only change is the way the temperature at $x = 1.0$ m is calculated. The boundary condition at $x = 1$ m is an ODE and, therefore, must also be discretized. In theory, this ODE can be discretized using *forward*, *backward* or *centered difference* (Eqs. 7.9, 7.10, or 7.11, respectively). As mentioned earlier, *centered difference* is more accurate; therefore, we will try to adopt it in this example. Discretization of Eq. (7.57) (at the position $x = 1$ m) yields:

$$\frac{T_{1.2,j} - T_{0.8,j}}{2\Delta x} = -\frac{h}{k}(T_{1.0,j} - T_{env}) \tag{7.58}$$

or

$$T_{1.2,j} = T_{0.8,j} - \frac{2\Delta x h}{k}(T_{1.0,j} - T_{env}) \tag{7.59}$$

Observe that the temperature at $x = 1.2$ m ($T_{1.2}$) is needed to obtain the temperature at $x = 1.0$ m (T_1) at the boundary. Since $T_{1.2}$ does not exist, we will consider an imaginary point at 1.2 m, and perform the calculus considering it. If the approach presented in Sect. 7.3 is adopted, Eq. (7.16) (rewritten below) can be used for all discretization points, including $x = 1$ m.

$$T_{i,j+1} = Fo\left(T_{i-1,j} - 2T_{i,j} + T_{i+1,j}\right) + T_{i,j} \tag{7.16}$$

$$\text{in which:} \quad Fo = \frac{k}{\rho c_p} \frac{\Delta t}{(\Delta x)^2}$$

At $x = 1.0$ m, Eq. 7.16 becomes:

$$T_{1.0,j+1} = Fo(T_{0.8,j} - 2T_{1.0,j} + T_{1.2,j}) + T_{1.0,j} \tag{7.60}$$

Observe that Eq. (7.60) presents the temperature at the imaginary point $T_{1.2,j}$. Substituting (7.59) in (7.60) and rearranging, yields:

$$T_{1.0,j+1} = T_{1.0,j} + 2Fo\left(T_{0.8,j} - T_{1.0,j}\left(1 + \Delta x \frac{h}{k}\right) + \Delta x \frac{h}{k} T_{env}\right) \tag{7.61}$$

To solve Eq. (4.15) with the boundary condition 7.57, the spreadsheet in Fig. 7.2 can be used again, but the cell *I15* must contain Eq. (7.61) instead of the value *30*.

Alternatively, the approach presented in Sect. 7.4.2 can be used; however, in addition to Eqs. (7.34), (7.35), (7.36) and (7.37), an ODE at $x = 1$ m (Eq. 7.62) is needed to represent the variation in temperature at this point over time:

$$\text{At } x = 1.0: \qquad \frac{dT_{1.0}}{dt} = \frac{k}{\rho c_p}\left(\frac{T_{0.8} - 2T_{1.0} + T_{1.2}}{(0.2)^2}\right) \tag{7.62}$$

As was done before, the discretized boundary condition at $x = 1$ (Eq. 7.57) generates the Eq. (7.63) for the temperature at the imaginary point $x = 1.2$ (differently from Eq. (7.59), herein the index j is not used because ODEs will be solved numerically over time latter on):

$$T_{1.2} = T_{0.8} - \frac{2\Delta x h}{k}(T_{1.0} - T_{env}) \tag{7.63}$$

Substituting Eq. (7.63) in (7.62) and rearranging, an ODE for $x = 1$ m is obtained (Eq. 7.64), which must be solved simultaneously with Eqs. (7.34), (7.35), (7.36) and (7.37):

$$\frac{dT_{1.0}}{dt} = \frac{2k}{\rho c_p (0.2)^2}\left(T_{0.8} - T_{1.0}\left(1 + \Delta x \frac{h}{k}\right) + \Delta x \frac{h}{k} T_{env}\right) \tag{7.64}$$

The spreadsheet presented in Fig. 7.7 can be used again, but substituting the cell *I15* with Eq. (7.64). The parameters ρ, c_p, and k are the ones shown in Fig. 7.7, and the heat coefficient (h) considered in this case is 300 J/min m^2 °C.

As mentioned earlier, *centered difference* was adopted in this example to discretize the boundary condition, because of its higher accuracy. To better visualize how *centered difference* is more precise, let us suppose that *backward difference* was used instead of *centered difference*. If this is the case, the discretization of the boundary condition (7.57) would yield:

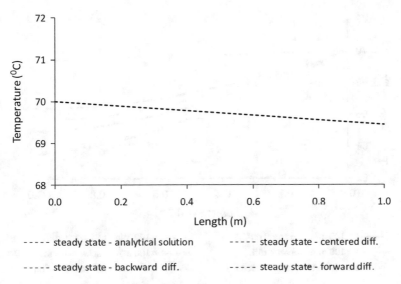

Fig. 7.10 Comparison between analytical and finite difference solutions in a steady state, applying centered, backward, and forward difference to discretize the boundary condition Eq. (7.57)

$$T_{1.0} = T_{0.8} - \frac{\Delta x\,h}{k}(T_{1.0} - T_{env}) \qquad (7.65)$$

Equation (7.65) can be rearranged to explicitly express $T_{1.0}$:

$$T_{1.0} = \frac{\frac{\Delta x\,h}{k}T_{env} + T_{0.8}}{\frac{\Delta x\,h}{k} + 1} \qquad (7.66)$$

If the approaches presented in Sects. 7.3 or 7.4.2 were used, the cell $I15$ in the spreadsheets in Figs. 7.2 or 7.7 should be replaced by the expression of Eq. (7.66). Analogously, *forward difference* could also be adopted.

Figure 7.10 compares *forward, centered* and *backward difference* applied to the boundary condition (7.57) when the system reaches a steady state. An analytical solution is also presented to better compare the results.

For this example, when a steady state is reached, the three ways of differencing present the same axial profiles, and they are equal to the analytical solution. On the other hand, before reaching the steady state, the type of differencing affects the results, as can be exemplified in Fig. 7.11 for time $= 100$ min. Even using smaller increment of x ($\Delta x = 0.1$), the accuracy obtained in *backward difference* (see Fig. 7.11) or in *forward difference* (not shown) is lower than the accuracy obtained in *centered difference* for $\Delta x = 0.2$. Observe that the curve representing the analytical solution for time $= 100$ min in Fig. 7.11 (see the equation for analytical solution in Appendix 7.2) matches the numerical solution using *centered difference* and $\Delta x = 0.2$ to represent the first derivative of the boundary condition (Eq. 7.57).

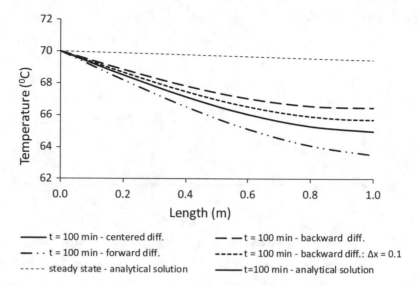

Fig. 7.11 Comparison between analytical and finite difference solutions before reaching a steady state (time $= 100$ min), applying centered, backward, and forward difference to discretize the boundary condition (Eq. 7.57). The analytical solution at the steady state (shown in Fig. 7.10) is also presented

The numerical procedures presented in this chapter can be used for different problems in chemical engineering involving PDEs, even if more independent variables are considered.

Proposed Problems

7.1) Imagine heat conduction in a cube with sides measuring 0.6 m. The cube is made of a metal with thermal conductivity k equal to 398 W/(m K). The initial temperature of the cube is 20 °C, but this temperature starts changing over time because all faces of the cube are kept at constant temperatures as depicted below:

Face	Superior	Inferior	Right	Left	Front	Back
T (°C)	200	30	50	150	100	80

a) Find a PDE that represents the temperature variation along the three coordinates (x, y, and z) and over time.
b) Consider $\Delta x = \Delta y = \Delta z = 0.2$ m and draw the cube (it is like a Rubik's cube). Discretize in x, y, and z and find eight ODEs representing the internal temperature variation over time.
c) Solve the system of ODEs using the RK4 method and plot the curves.

7.2) Imagine a beaker of radius (R) 1 cm and 5 cm high (L), open at the top with only air inside it, as per Example 4.7. A certain gas A at a concentration $C_A = 1$ kmol/m^3 starts flowing around the beaker. There is diffusion of A inside the beaker, but at the bottom, C_A always remains equal to zero. Assume that the mass diffusivity of A in the air is $D_A = 0.018$ m^2/h. The equation that represents this system is presented by Eq. (4.23) rewritten below:

$$\frac{\partial C_A}{\partial t} = D_A \frac{\partial^2 C_A}{\partial x^2} \tag{4.23}$$

The initial and boundary conditions are:

At $t = 0$ h, $C_A = 0$ kmol/m^3, for $0 \leq x \leq 0.05$ m
At $x = 0$ m, $C_A = 0$ kmol/m^3, for $t > 0$ h
At $x = 0.05$ m, $C_A = 1$ kmol/m^3, for $t > 0$ h

Use the finite difference method to obtain the axial profile of the concentration of A inside the beaker over time.

7.3) Consider Example 4.8 about two concentric cylinders. A copper cylinder of length 1 m and radius 0.1 m at a constant temperature of 80 °C is coated with an annulus made of aluminum, initially at 50 °C. The total radius of the concentric cylinders (copper plus aluminum) is 0.3 m. The environment temperature is constant and equal to 25 °C. Although the aluminum exchanges heat with the environment, the two ends of the two concentric cylinders are insulated. The model for this system is represented by Eq. (4.27), rewritten below:

$$\frac{\partial^2 T}{\partial r^2} + \frac{1}{r}\frac{\partial T}{\partial r} = \frac{\rho c_p}{k}\frac{\partial T}{\partial t} \tag{4.27}$$

The initial and boundary conditions are:

$t = 0$, $T = 50$ °C, for $0.1 \leq r \leq 0.3$
$t > 0$, $r = 0.1$, $T = 80$ °C
$t > 0$, $r = 0.3$, $\frac{dT}{dr} = -\frac{h}{k}(T - 25)$

Use the finite difference method to find profiles of the temperature along the radius and over time until a steady state is reached. Consider $k = 180$ W/mK, $h = 100$ W/m^2 K, $c_p = 0.91$ KJ/kg K and $\rho = 2.7$ g/cm^3.

Appendix 7.1

Figures A.7.1, A.7.2, and A.7.3 show the VBA code developed to solve the ODE system generated in Sect. 7.4.3 (Eqs. 7.47, 7.48, 7.49, 7.50, 7.51, 7.52, 7.53, 7.54, 7.55 and 7.56). Observe that the code is almost identical to the one presented in Chap. 6 (Figs. 6.18b, 6.17, and 6.20). The function *Derivative* is changed to account for the new system of ODEs. The function *RungeKutta4* is identical, except for the dimensions of variables $k1, k2, k3, k4$, and $yrun$, which have changed from 5 to 10, to account for the 10 ODEs. The main program *RK4* is modified only in the

```
Sub RK4()

Dim dydx(10) As Double        ⟸      dimension
Dim y(10) As Double

xi = 0          '(hour)
xf = 0.028      '(hour ~ 100 sec)
y(1) = 0        '(kmol/m3)
y(2) = 0        '(kmol/m3)
y(3) = 0        '(kmol/m3)
y(4) = 0        '(kmol/m3)         Initial conditions
y(5) = 0        '(kmol/m3)
y(6) = 200      '(C)
y(7) = 200      '(C)
y(8) = 200      '(C)
y(9) = 200      '(C)
y(10) = 200     '(C)

dx = 0.00028 '(hour ~ 1 sec)  ⟸    step

x = xi

Cells(1, 1) = "t"
Cells(1, 2) = "Y1"
Cells(1, 3) = "Y2"
Cells(1, 4) = "Y3"
Cells(1, 5) = "Y4"
Cells(1, 6) = "Y5"
Cells(1, 7) = "Y6"
Cells(1, 8) = "Y7"
Cells(1, 9) = "Y8"          ⟸     dimension
Cells(1, 10) = "Y9"
Cells(1, 11) = "Y10"

i = 2

ny = 10

Do While x <= xf

    Cells(i, 1) = x
    For k = 1 To ny
        Cells(i, k + 1) = y(k)
    Next

    Call RungeKutta4(x, y, dydx, ny, dx)

    i = i + 1

Loop


End Sub
```

Fig. A.7.1 The main program in Visual Basic for Applications (VBA) code to solve an ordinary differential equation (ODE) system (Eqs. 7.47, 7.48, 7.49, 7.50, 7.51, 7.52, 7.53, 7.54, 7.55 and 7.56) using the fourth-order Runge–Kutta (RK4) method

```
Function RungeKutta4(x, y, dydx, ny, dx)

Dim k1(10) As Double
Dim k2(10) As Double
Dim k3(10) As Double           <]  dimension
Dim k4(10) As Double
Dim ytran(10) As Double

'Calculate K1 for all ODEs
Call Derivative(x, y, dydx)
For i = 1 To ny
    k1(i) = dydx(i)
Next
'Define x and y to calculate K2
For i = 1 To ny
    ytran(i) = y(i) + k1(i) * 0.5 * dx
Next
xtran = x + 0.5 * dx

'Calculate K2 for all ODEs
Call Derivative(xtran, ytran, dydx)
For i = 1 To ny
    k2(i) = dydx(i)
Next
'Define x and y to calculate K3
For i = 1 To ny
    ytran(i) = y(i) + k1(i) * 0.5 * dx
Next

'Calculate K3 for all ODEs
Call Derivative(xtran, ytran, dydx)
For i = 1 To ny
    k3(i) = dydx(i)
Next
'Define x and y to calculate K4
For i = 1 To ny
    ytran(i) = y(i) + k1(i) * 0.5 * dx
Next
xtran = x + dx

'Calculate K4 for all ODEs
Call Derivative(xtran, ytran, dydx)
For i = 1 To ny
    k4(i) = dydx(i)
Next

'Uptade dependent variables
For i = 1 To ny
    y(i) = y(i) + (1 / 6) * (k1(i) + 2 * k2(i) + 2 * k3(i) + k4(i)) * dx
Next

'Update the independent variable
x = x + dx
End Function
```

Fig. A.7.2 Function of the fourth-order Runge Kutta (RK4) method called in the main program (Fig. A.7.1)

```
Function Derivative(x, y, dydx)

    Q = 32                          '(m3/hour)
    A = (3.1416 * (0.2425) ^ 2) / 4  '(m2)
    Ca0 = 18.75                     '(kmol/m3)
    T0 = 200                        '(C)
    DeltaX = 0.6                    '(m)
    DHR = -15000                    '(kJ/kmol)
    rhoCp = 787.5                   '(kJ/m3 K)

    dydx(1) = ((-Q / A) * (y(1) - Ca0) / DeltaX) - (110 + 0.8 * (y(6) - 200)) * y(1)
    dydx(2) = ((-Q / A) * (y(2) - y(1)) / DeltaX) - (110 + 0.8 * (y(7) - 200)) * y(2)
    dydx(3) = ((-Q / A) * (y(3) - y(2)) / DeltaX) - (110 + 0.8 * (y(8) - 200)) * y(3)
    dydx(4) = ((-Q / A) * (y(4) - y(3)) / DeltaX) - (110 + 0.8 * (y(9) - 200)) * y(4)
    dydx(5) = ((-Q / A) * (y(5) - y(4)) / DeltaX) - (110 + 0.8 * (y(10) - 200)) * y(5)
    dydx(6) = ((-Q / A) * (y(6) - T0) / DeltaX) + (110 + 0.8 * (y(6) - 200)) * y(1) * ((-DHR) / (rhoCp))
    dydx(7) = ((-Q / A) * (y(7) - y(6)) / DeltaX) + (110 + 0.8 * (y(7) - 200)) * y(2) * ((-DHR) / (rhoCp))
    dydx(8) = ((-Q / A) * (y(8) - y(7)) / DeltaX) + (110 + 0.8 * (y(8) - 200)) * y(3) * ((-DHR) / (rhoCp))
    dydx(9) = ((-Q / A) * (y(9) - y(8)) / DeltaX) + (110 + 0.8 * (y(9) - 200)) * y(4) * ((-DHR) / (rhoCp))
    dydx(10) = ((-Q / A) * (y(10) - y(9)) / DeltaX) + (110 + 0.8 * (y(10) - 200)) * y(5) * ((-DHR) / (rhoCp))

End Function
```

Fig. A.7.3 The function *Derivative* with ten ordinary differential equations (ODEs) (Eqs. 7.47, 7.48, 7.49, 7.50, 7.51, 7.52, 7.53, 7.54, 7.55 and 7.56) called in the function *RungeKutta4*

dimension of the variables and values for the initial conditions and integration step, as highlighted in Fig. A.7.1.

Appendix 7.2

Figure 7.11 shows a curve for the analytical solution of Eq. (4.15), rewritten below:

$$\frac{\partial T}{\partial t} = \frac{k}{\rho c_p} \frac{\partial^2 T}{\partial x^2} \tag{4.15}$$

At $t = 0$ h, $T(x,0) = 50\,°C$, for $0 \le L \le 1$ m

At $x = 0$ m, $T(0,t) = 70\,°C$, for $t > 0$ h

At $x = 1$ m, $\dfrac{dT}{dx} = -\dfrac{h}{k}(T - T_{env})$, for $t > 0$ h $\tag{7.57}$

The analytical solution of Eq. (4.15) can be obtained using the Fourier method, to yield:

$$T = T(0,t) + \left(\frac{h}{h+k}\right)(T_{env} - T(0,t)x)$$

$$+ \sum_{n=1}^{\infty} \frac{2}{\left(1 + \frac{k}{h}(\cos\beta_n)^2\right)} \left(\frac{T(x,0) - T(0,t)}{\beta_n} - (T(x,0) - T_{env})\frac{\cos\beta_n}{\beta_n}\right)$$

$$\times \exp\left(-\frac{k}{\rho\,cp}\beta_n^2\,t\right)\sin(\beta_n x)$$

in which β_n are the solutions of:

$$\tan \beta_n + \frac{k}{h}\beta_n = 0 \quad \text{For n} = 1,2,3,\ldots$$

References

Carnahan, B., Luther, H.A., Wilkers, J.O.: Numerical Applied Methods. Wiley, New York (1969)
Chapra, C.C., Canale, R.P.: Numerical Methods for Engineers, 5th edn. Mc Graw Hill, New York (2005)
Davis, M.E.: Numerical Methods and Modeling for Chemical Engineers. Wiley, New York (1984)
Hill, C.G., Root, T.W.: Introduction to Chemical Engineering Kinetics and Reactor Design, 2nd edn. Wiley, New York (2014)

Errata to:
A Step by Step Approach to the Modeling of Chemical Engineering Processes – Using Excel for Simulation

Liliane Maria Ferrareso Lona

Errata to:
L.M.F. Lona, *A Step by Step Approach to the Modeling of Chemical Engineering Processes,* **https://doi.org/10.1007/978-3-319-66047-9**

In the original version of the book published, there were some errors in equations of chapters 4, 5, 6 and 7 which have been updated now in this version, as per author's request.

The updated online versions of these chapters can be found at
https://doi.org/10.1007/978-3-319-66047-9_4
https://doi.org/10.1007/978-3-319-66047-9_5
https://doi.org/10.1007/978-3-319-66047-9_6
https://doi.org/10.1007/978-3-319-66047-9_7

The updated online version of this book can be found at
https://doi.org/10.1007/978-3-319-66047-9

Index

© Springer International Publishing AG 2018
L.M.F. Lona, *A Step by Step Approach to the Modeling of Chemical Engineering Processes*, https://doi.org/10.1007/978-3-319-66047-9

171

Printed in the United States
By Bookmasters